1st Edition Reviews

"It simply doesn't get any better than this. If you buy one book to help you find a career in renewable energy, make it this one. Hands down, the best green job book on the market today."
GreenMuze staff

"This book is a gem! For those who are searching for a rewarding career in renewable energy, this is the place to start. It is chock full of important information and resources that will guide you on your journey. I highly recommend it."
Susan T. Schleith, Education Coordinator, Florida Solar Energy Center

"The clean energy economy is booming across the United States and around the world. *Careers in Renewable Energy* is a sophisticated—and user-friendly—guide for finding meaningful job opportunities in the renewable energy sector."
Bill Richardson, Governor of New Mexico

"This book is an enjoyable read, succinct yet complete. It's a must-have for anyone considering a new or different career in renewable energy, or anyone advising them! And if you weren't already considering a new career? You might after you read THIS book!"
Dr. Cortney Martin, Virginia Tech

"Edifying and accessible, this volume will be a welcome addition for career centers and environmental collections."
School Library Journal

"*Careers in Renewable Energy* will open your students' eyes to the many possibilities that already exist, as well as ones that are bound to unfold in the years ahead."
Peter Vogt, Campus Career Counselor publication

"The green industries are growing at an unforeseen rate, and are concerned with meeting their growth potential. This book is the perfect guide to gain a foothold in these dynamic industries."
Scott Sklar, President – The Stella Group, Ltd.

"This useful book is long overdue. More and more often, I receive calls from people across the country wanting to know how to get into solar; what kind of training and experience is required, and where to get it. Finally a resource that definitively answers these questions and much more."
Jonathan Hill, Solar Applications Engineer – Sierra Solar Systems

Careers in Renewable Energy

Your World. Your Future.

UPDATED 2ND EDITION

Gregory McNamee

100% solar & wind powered since 1999
PIXYJACK PRESS INC.

CAREERS IN RENEWABLE ENERGY

Your World. Your Future.

Updated 2nd Edition

Copyright © 2014 by Gregory Mcnamee

No part of this book may be reproduced, stored in a retrieval system or transmitted in any form, or by any means, electronic, mechanical, photo-copying, recording or otherwise, without prior written permission of the publisher, except by a reviewer, who may quote brief passages in review.

Published by PixyJack Press, Inc.
PO Box 149, Masonville, CO 80541 USA

Second Edition
print ISBN 978-1-936555-52-9
Kindle ISBN 978-1-936555-53-6
epub ISBN 978-1-936555-54-3

Front cover photo by Meg Takamura; back cover car photo by NREL.

First Edition © 2008 (ISBN 978-0-9773724-3-0)

Library of Congress Cataloging-in-Publication Data

Mcnamee, Gregory, author.
  Careers in renewable energy : your world, your future / by Gregory Mcnamee. -- Updated 2nd edition.
      pages cm
  Summary: "Vocational guidance for students and adults seeking jobs in renewable energy, including solar and wind energy, geothermal, hydropower, bioenergy, green building / energy management, hydrogen energy, green transportation, and energy education and economics"-- Provided by publisher.
  Includes bibliographical references and index.
  ISBN 978-1-936555-52-9 -- ISBN 978-1-936555-53-6 (Kindle) -- ISBN 978-1-936555-54-3 (epub)
  1. Clean energy industries--Vocational guidance. 2. Renewable energy sources. 3. Energy industries--Vocational guidance. 4. Environmental sciences--Vocational guidance. I. Title.
  HD9502.5.C542M35 2014
  333.79'4023--dc23
                        2013048664

Contents

Introduction

In 1987, running a side-by-side refrigerator used about 950 kilowatt hours of electricity and cost about $150 a year. A comparable refrigerator now uses half the electricity, to the benefit of both environment and wallet.

In 1975, there were 3,775,427 automobiles on the streets of metropolitan Los Angeles. As of 2012 there were more than 5,800,000. Air-pollution levels, however, have fallen roughly in half in the intervening decades, and an increasing number of the automobiles now on the road are hybrids or rely on biodiesel or other renewable fuels.

As long ago as the 1890s, a third of the homes in southern California were equipped with solar water heaters. By the 1920s, the number had

fallen to almost none. The number began to rise again in the 1970s. And so did the use of solar power generally. As of 2007, along with a substantial number of homes with solar thermal hot-water systems installed, fully half a million homes in Southern California were receiving direct solar power, either from solar electricity plants or from rooftop photovoltaic panels. And by 2030, the Department of Interior projects, solar facilities will occupy 154,000 acres of federal land in the state, providing the lion's share of energy to consumers in the Golden State.

California is not alone: across North America, Europe, and Asia, businesspeople, municipalities, and other governments are working to transform the world's energy mix.

These are changing times indeed. Students, workers, consumers, businesspeople, and government officials are increasingly aware that humans have been placing terrific stresses on the environment, as global warming, water shortages, mass extinctions of plant and animal species, and the dwindling supply of fossil fuels attest. In times of trouble, uncertainty, and even crisis, as entrepreneurs will tell you, there are opportunities. And, as the preceding examples suggest, there are reasons to be optimistic, for a rising generation of students, workers, consumers, businesspeople, and government officials—people, in short, from every walk of life and from every corner of the world—is more and more committed to the idea that we can bring the most important power of all to bear on the problem of where our power comes from—that being brainpower.

Look at it this way: a championship bicycle racer burns as much energy as a handheld hair dryer. So does a cheetah running at full clip. Our brains use about as much as a refrigerator bulb, whether we're thinking hard or barely sentient. It doesn't tax us too much to think, hard and long, about what we can do to lessen our footprints on the land. We have a big problem at the outset: the world's people use nearly 80 million barrels of oil a day, and there is much work to do to wean ourselves from our addiction to irreplaceable fossil fuels.

This book is about the many ways in which we can do that by working with "soft," renewable forms of energy: power derived from the wind, the sun, the sea, the earth, and biological sources, power that causes little or no damage and that can constantly be replaced. Getting there requires a lot of brainpower—and some elbow grease, too.

Jobs in many of those renewable forms of energy were once few and far between. These days, though, companies are going lean, clean, and green. In every profession, in every walk of life, individual consumers, institutions, and corporations are looking to do things better and smarter in the work of producing and using energy and in lessening our dependence on fuels that will one day run out—perhaps sooner than we think.

As high school and college graduates will quickly discover, all other things being equal, there is no better time to learn environmentally responsible habits and put them to work in the marketplace. "Green jobs" are becoming ever more common, and demand for green things is on the rise, even as increased calls for energy independence are in the air. New fields are emerging as a result: a few years ago, there was no such thing as someone who brokered carbon trades, for instance, and very few builders and buyers who used terms such as "green building."

The fastest-growing professions within the so-called green-jobs realm, according to Kevin Doyle, president of the Boston-based consulting company Green Economy, are environmental engineers, hydrologists, environmental-health scientists, and urban and regional planners. Employers with such positions to fill look for students who major in engineering, mathematics, earth sciences, environmental studies, public policy, and economics—but also who come from liberal arts or science backgrounds generally, as long as they can think through a problem and communicate solutions to it.

The green-job phenomenon is also an important aspect of the trades these days, and skilled workers are needed to build, install, operate, and maintain the workings of green energy: rooftop solar panels, geothermal piping, wind turbines, fuel cells, wave rotors, radiant flooring, the list goes on and on.

U.S. Energy Consumption per Person

In 1949, energy use per person was 215 million Btu. In 2011, it was 312 million Btu.

SOURCE: WWW.EIA.GOV

Going green involves all of us. The world is in trouble. Or, perhaps better put, we're in trouble in the world. Getting us out of our current fix is going to take a lot of thought, a lot of work—and a lot of energy.

Careers in Renewable Energy

In this book, we'll look closely at dozens of careers within the broad renewable-energy field. Where possible, we'll use firm figures and talk dollars and cents. For instance, according to the *Environmental Business Journal*, the green industry in the United States in 2005 was about $265 billion, employing 1.6 million people. Green businesses had been growing at a rate of about 5 percent annually since then until the Great Recession hit in 2007. Had the job-growth trajectory continued at a normal pace, 85,000 new jobs would have opened up each year, and now that the recession is abating, job growth is trending toward this number. Some of these jobs are new. Others are new wrinkles on old fields such as the law, architecture, or business administration. Some are quite old: we need farmers to produce the makings of biofuel, for instance, and smooth-talking politicians to convince their colleagues to part with research money and provide grants and tax incentives so that homeowners and businesses can retrofit to take advantage of the energy-efficient, clean technologies of today and tomorrow.

And, as always, we need inventors, entrepreneurs, and visionaries to bring us new technologies and spread the word about them—the same people who, once upon a time, brought us electricity. We hope that some of the information you'll find in the pages ahead will inspire those kinds of people to get going on the big solutions we need to our big problems—and the big opportunities they occasion.

Knowledge Is Power

We also focus here on some of the places that offer training for entering the world of renewable energy: trade schools, community colleges, four-year colleges and universities, and graduate programs, along with professional organizations that sponsor workshops, conferences, and other continuing-education opportunities.

Because so many aspects of the green-energy field are new, there are not always abundant programs to match them. At MIT, for instance, that world-renowned school, there is no major as such in "renewable energy," though the interested student will find plenty of fields, from engineering to physics to business, that deeply concern themselves with renewable-energy matters. With guidance, a student could even design his or her own program with some of the courses scattered about in several departments, such as:

- Alternate Energy Sources
- Ecology I: The Earth System
- Ecology II: Engineering for Sustainability
- Fundamentals of Advanced Energy Conversion
- Material Science
- Mechanical Engineering
- Multiscale Analysis of Advanced Energy Processes
- Ocean Wave Interaction with Ships / Offshore Energy Systems
- Sustainable Energy

Not everyone can go to MIT, of course. Not everyone wants to, and not everyone needs to. Many community colleges offer training in alternative energy technology, for instance, in which students, one catalog tells us, "are introduced to design issues associated with home construction, com-

Green transportation is just one of the sectors where jobs can be found. Photo: Hydrogen fuel cell bus in Amsterdam; courtesy of Shell Hydrogen.

munity development and passive solar design." That particular program is made up of required courses that are as rigorous as MIT's, at first glance:

- Business English
- Building the Human Environment
- Construction Materials and Equipment Safety
- Building Construction Methods I
- Solar Home Design
- Intermediate Algebra

Add another year's study to that made up of the following courses, and you'll earn an advanced certificate:

- Introduction to Computer Information Systems
- Technical Drafting & CAD Fundamentals
- Micro Economics Principles
- Blueprint Reading
- Building Construction Methods II
- Photovoltaics and Wind Power
- Innovative and Alternative Building Techniques
- Technical Problem Solving

That's quite a full plate. Any program of study is going to be. The pay-offs are immediate, though, not just in the sense of preparing a student to enter the renewable-energy field, but also in rewarding that student for his or her labors and preparation.

Of course, those with college training have greater earning power, generally speaking, than those without it. According to the University of Wisconsin Engineering Career Service (*https://ecs.engr.wisc.edu*), for

Oil for Transportation / Coal for Electricity

In 2011, transportation got 93% of its energy from petroleum, 3% from natural gas, and 4% from renewable energy.

In 2011, 92% of coal's energy went to electric power for the public, plus 8% to industrial.

SOURCE: WWW.EIA.GOV

instance, in 2013, a man or woman holding a bachelor's degree in civil and environmental engineering and entering the profession earned a median of $54,223 at the outset; a degree in industrial engineering commanded $61,505; and one in mechanical engineering brought slightly less, $60,798. Add a master's degree to the qualifications, and those figures rose to $55,096, $65,710, and $77,129 respectively. Add a Ph.D., and the entry-level pay for a mechanical engineer climbed to $86,125. The school did not place a civil and environmental or an industrial engineer in jobs 2013, so no figures are available—but we can assume they would show a similar spread.

Knowledge is power. And, in the renewable-energy field, knowledge is money, too. In this book, we hope to provide access to all the sources you'll need in order to make the career in renewable energy that's just right for you. ❖

Where To Study Renewable Energy

Scott Sklar is president of The Stella Group, Ltd. optimizes high-value energy efficiency and renewable energy technologies for facilities or buildings, blends financing, and insures applications are standardized, modular, and web-enabled. Sklar teaches two unique interdisciplinary courses on sustainable energy at The George Washington University (GWU) and is an affiliated professor at the international sustainable graduate university (CATIE) in Costa Rica. In December 2012, Sklar was appointed by Acting Secretary Rebecca Blank to serve on as Vice Chairman of the Department of Commerce Renewable Energy and Energy Efficiency Advisory Committee, term ending June 2014. His co-authored book, *A Consumer Guide to Solar Energy,* is in its third printing, and he lives in a solar home and has solar on both his office buildings.

We asked him to name the ten schools he would recommend to a student interested in pursuing a career in renewable energy, and these are his picks:

- Arizona State University
- Bradley University
- Colorado State University
- George Washington University
- Massachusetts Institute of Technology (MIT)
- North Carolina State University
- Syracuse University
- University of California (Merced)
- University of Central Florida
- University of Nevada (Reno)

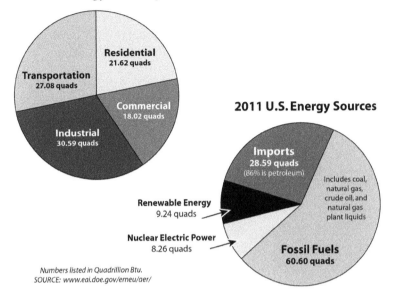

2011 U.S. Energy Consumption

Residential
21.62 quads

Transportation
27.08 quads

Commercial
18.02 quads

Industrial
30.59 quads

2011 U.S. Energy Sources

Imports
28.59 quads
(86% is petroleum)

Includes coal,
natural gas,
crude oil, and
natural gas
plant liquids

Renewable Energy
9.24 quads

Nuclear Electric Power
8.26 quads

Fossil Fuels
60.60 quads

Numbers listed in Quadrillion Btu.
SOURCE: www.eai.doe.gov/emeu/aer/

. . . .

Life-Cycle Engineering

A quiet revolution is underway in engineering classrooms in Canadian universities, prompted by a climate change crisis that has unleased a torrent of new green technologies. It's called Life-Cycle Engineering. And, simple put, it means engineers will design products that can be manufactured while leaving a minimal environmental footprint. LCE involves estimating the environmental costs and social benefits at every stage of the life cycle—starting with raw material extraction, through product processing, transport, distribution and finally disposal. More importantly, this "cradle-to-grave" approach to engineering allows the engineer to sift through the hype and critically analyze the true impact of so-called green technologies – their promise and their pitfalls.

Source: Sumitra Rajagopalan, thestar.com

. . . .

Google Goes Green in a Big Way

Google wanted to find a way to reduce energy costs at its Mountain View "Googleplex" as well as make a statement in support of clean energy. E1 Solutions engineered a way to mount 10,330 Sharp solar modules on every available rooftop plus the carports.

By installing the largest solar power system ever installed at a single corporate campus, the 1.6 megawatt-system will save Google more than $393,000 annually in energy costs, and the system will pay for itself in approximately 7.5 years. And CO_2 emissions are reduced by 3.6 million pounds per year!

Source: E1 Solutions, www.eispv.com

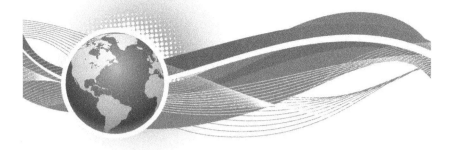

Solar Energy

In 2007, solar energy provided less than one percent of the power used in the North American grid. In 2027, by some estimates, as much as a quarter of America's energy mix may come from solar sources. There are obstacles to be overcome, some technological, some institutional—and those obstacles mean opportunity at many levels, from the hunger of an audience looking for better and more efficient ways to capture the sun's energy to the need for skilled workers in installing, maintaining, and operating solar-energy systems. These systems may be massive, such as the great solar plant at Kramer Junction, California (for more on this sprawling facility, see *www.clui.org/ludb/site/kramer-junction-solar-electric-generating-station*), or small, such as a simple set of rooftop-mounted solar panels capable of providing some or all of a single home's electrical power

Photo: Neville Williams

needs. Whatever the case, these systems require well-trained, smart people to develop and improve them and to keep them running efficiently.

Solar energy works by converting energy from the sun, in the form of light, into electricity. This is accomplished by several means. One is to use photovoltaic (PV) cells made up of semi-conductors. Rooftop solar panels, for example, are typically made of multiple cells encased in weather-resistant housing, and these cells convert sunlight into direct current. (Other kinds of panels are used to heat water directly; these solar thermal systems do not generate electricity but are used to heat homes and pools.) Another is to channel sunlight onto fluid to cause it to boil and thereby spin a turbine; at the Kramer Junction plant, for instance, a field of highly polished mirrors concentrates reflected light onto a synthetic oil that passes through water to produce clean-burning steam that in turn drives a turbine, powering thousands of nearby houses. Still another is to use engines that are directly powered by solar heat, such as the Stirling engine, in which gas heated by the sun pushes a piston that in turn powers a generator; as the gas leaves the heat source it cools, and eventually it returns through the system to go back into the cycle anew.

The future will doubtless bring new means of using the sun's energy to make power that is directly useful to us. It is up to physicists, material scientists, engineers, and inspired inventors from inside and outside academia to come up with those new means. It is up to technicians and workers of many kinds to get the results of that engineering online and available to us, and to keep it running. Many kinds of thought and experience are needed in the field—and many kinds of careers in solar energy await people from just about every background imaginable.

A growing population requires homes, shops, and workplaces, and this in turn opens many prospects for those interested in careers in installing and maintaining energy systems. Industry estimates indicate continued strong growth in solar energy jobs worldwide, with new jobs being created in both the residential and nonresidential sectors: marketing and installing solar photovoltaic and thermal systems for homes, businesses, schools and hospitals, and other facilities, for instance, and in building and maintaining utility-scale installations such as the one at Kramer Junction.

Other countries make broad use of solar energy, with Germany representing a case in point: in 1998, the German solar industry employed only about 1,500 people, but in 2020 the number of workers in solar and other renewable forms of energy will top 500,000, and it is expected to reach nearly 750,000 by 2030. Such a dramatic transformation in energy regimes and the jobs to service the new energy economy requires both political will and money, of course, which Germany has backed up by an ambitious program of federal loans and other economic incentives. Other European countries are projecting a similar growth rate: "By the time the generation born today reaches adulthood in 2020," the European Photovoltaic Industry Association declares, "solar energy could easily provide energy to over a billion people globally and provide 2.3 million full-time jobs."

For the moment, the leaders in solar energy in the United States are located in areas where, not surprisingly, the sun shines brightly throughout the year. Florida is home to the largest solar photovoltaic plant in North America, the DeSoto Next Generation Solar Energy Center, which went online in 2009. California and Nevada, meanwhile, are leaders on the West Coast, and California enjoys the largest percentage of solar energy in the overall energy mix. But it pays to look around for opportunities wherever you are, which may pop up in unexpected corners.

In the most recent industry survey, for instance, most domestic shipments of PV modules went to five states: California, Florida, Arizona, Colorado, and—perhaps surprisingly—New Jersey. Meanwhile, according to *Renewable Energy World* magazine's 2012 jobs report, the leading states for employment in the solar field were, in order, California, Colorado, Arizona, Pennsylvania, New York, Florida, Oregon, Texas, New Jersey, and Massachusetts. Growth in the field accompanies a general shift, slow but sure, from small-scale installations to what industry insiders call "utility solar." As Roger Duncan, former general manager of the Texas firm Austin Energy, puts it, "We've gotten out of the kilowatt stage and the megawatts, and now we're in the hundreds of megawatts being developed. You'll see large growth in the solar industry over the next decade."

Solar Home Facts

A 2.2-kilowatt home solar PV system that consists of twelve 180-watt photovoltaic panels requires about 225 square feet of installation area, easily available on most roofs. It would generate approximately 2,800 kilowatt hours per year in a sunny climate; about half that in a cloudy climate.

Careers in Solar Energy

Within the solar-energy field, there are white-collar, blue-collar, and what might be called green-collar jobs, altogether making a mix of administrative, scientific, business, and technical skills. Where you land on the ladder largely depends on your interests and abilities, as well as on the course of education and specialized training you follow.

Administrative, Financial, and Nontechnical

Most jobs in the white-collar realm require, at a minimum, a degree from a four-year college. Many workers at the higher levels of utility administration, for example, hold degrees in business or economics. Chief executive and business-development officers usually have a solid education combined with broad professional experience of at least ten years' time; finance directors, account-ants, and corporate strategists have excellent number-crunching skills. In all white-collar levels of the energy business, good mathematical and communication skills are considered to be of great importance, since an executive has to have a grasp on business fundamentals and, ideally, should be able to communicate the vision of his or her company to the outside world and constituents within the organization.

Other nontechnical jobs are as numerous and varied as technical ones within the field. Companies large and small require marketing, public relations, and human-resources services, and many careers in the renewable-energy market will involve sales, or a combination of sales and technical skills. One Southern California employer ran an ad for an inside sales position, for example, that called for "strong leadership skills, a positive, can-do attitude, 2 to 5 years of experience in business, world-class communications and cus-

Consumers can cut their energy bills by up to 50 percent by investing in solar water-heating technologies.

tomer-service skills, the ability to work long hours, and strong math and computer skills." Added to this was the strong preference that the successful applicant hold a four-year college degree. The responsibilities of the job, the employer specified, were these:

Learn about the solar industry, current technology and the rebate programs in order to effectively respond to general inquiries from potential customers. Manage incoming new leads and general email communication. Qualify new leads by completing customer questionnaire. Add customers into customer database. Log all communications with customers in customer database. Assist in special projects as needed, perform research and maintain records. Coordinate mass mailing, conferences, and marketing efforts as required by program departments. Serve as backup office manager. Set appointments for site evaluations.

In a similar spirit, the Department of Energy's National Renewable Energy Laboratory (NREL) recently announced a job opening for a technical writer and editor, another nontechnical job that, of course, has many specialized requirements of its own. At a minimum, the posting specified:

Bachelor's degree in communications, journalism, or English or equivalent experience, plus five years of relevant work experience in a scientific or technical information environment. The selected candidate must have a good understanding of communications theory and its application and must be able to provide highly detailed editing of technical and outreach print publications and web sites. Selected candidate will conduct research and write technical and outreach material for a variety of audiences, including the general public, interested stakeholders, research partners, and policy makers. Demonstrate project management and business development skills and work on multiple communications projects simultaneously. Personal computer skills such as word-processing, Internet browser, and e-mail applications are required.

Research

Solar-energy research positions call on academic and business skills of many kinds. A look at the NREL home page (*www.nrel.gov*) quickly shows that a solid grounding in mathematics, physics, chemistry, and other sciences is essential to most such positions within the field.

Consider what sort of background might be required, for instance, to enter the following program, as described on NREL's site:

The U.S. Department of Energy (DOE) researches and develops a clean, large-scale solar thermal technology known as concentrating solar power (CSP). This research and development (R&D) focuses on three types of concentrating solar technologies: trough systems, dish/engine systems, and power towers. These solar technologies are used in CSP plants that use different kinds of mirror configurations to convert the sun's energy into high-temperature heat. The heat energy is then used to generate electricity in a steam generator.

Concentrating solar power's relatively low cost and ability to deliver power during periods of peak demand—when and where we need it—mean that it can be a major contributor to the nation's future needs for distributed sources of energy.

DOE's Solar Energy Technologies Program pursues concentrating solar power R&D to provide clean, reliable, affordable solar thermal electricity for the nation. The National Renewable Energy Laboratory (NREL) and Sandia National Laboratories work together as SunLab, a partnership developed by DOE to support R&D within the Concentrating Solar Program.

One indication might be NREL's requirements for an entry-level engineering position:

Bachelor's degree in engineering or closely related field, or equivalent relevant experience, plus a demonstrated understanding of relevant R&D. Fundamental knowledge of engineering practices, procedures, and concepts. Basic engineering abilities in practices and techniques; applies basic technical skills and methods to help analyze and solve problems. Demonstrated adequate interpersonal and communication (oral and written) skills. Must be computer literate including programming, interfacing, and software development and/or utilization.

At a far more advanced level leading to a senior research position, NREL advertised for a postdoctoral fellow in photovoltaics, the inner workings of solar panels:

A recent Ph.D. (less than three years), in Chemistry, Chemical Engineering, or Physics. Candidates must have: a strong background in ink-based materials, solution patterning, liquid processing and thin film deposition; strong scientific writing skills demonstrated by experience in writing quality scientific papers; strong communications skills; and, the ability to work collaboratively as part of a research team.

Similarly, Heliovolt, Inc., a company in Austin, Texas, specializing in the manufacture of thin, light solar installations, recently listed a job in thin-film circuit manufacture. The job calls for a master's or doctoral degree in physics or applied math, as well as "experience designing and modeling thin film circuits, strong mathematical background, experience using modeling and simulation tools (SPICE, Mathematica, etc.), and a general knowledge of thin film fabrication processes." This thin-film technology is extremely promising, and in 2012 3M introduced a solar film for windows that generated electricity affordably while making those windows shatterproof, a double advantage for homeowners.

Research and theoretical skills, to say nothing of practical skills, are hard to attain. Without trained workers and thinkers, the renewable-energy field cannot advance. Those who are best prepared, through broad study and hands-on experience alike, will be the most successful.

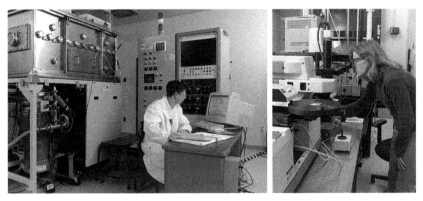

Left: Research is being conducted on thin film solar cells. Right: An NREL researcher uses an infrared microscope to examine novel PV materials. Photos: Institute of Energy Conversion; Mike Linenberger, NREL

Technical

By every projection, despite the long economic downturn, it's a seller's market for technicians and other crafts workers well trained in installing and maintaining solar-energy systems. Demand is expected to exceed supply considerably over at least the next decade, and though the entry-level qualifications are rigorous, the rewards are substantial.

Here, for example, is an excerpt from a help-wanted listing posted by a solar-energy contractor in southern California, calling for a solar systems installer. The desired qualifications, the employer says, are these:

- Strong work ethic, self-motivated, organized and a can-do attitude.
- Experience in mechanical installations, general construction, and ladder work.
- Basic understanding/experience with electrical wiring of AC and DC systems (preferred).
- 1–2 years of construction background is preferred.
- Experience with all types of hand-held and power tools.
- Experience with small machinery (trenchers, bobcats, forklifts).
- Experience working with all types of building materials; various roof types, stucco, wood, concrete, Uni-strut, roofing etc.
- Team player who listens, learns, and actively communicates.
- Visual thinker good at problem solving and implementing ideas.
- Knowledge of the renewable-energy marketplace, technology, and industry.
- Desire to learn and master all aspects of installing solar photovoltaic systems.
- High-school diploma, two-year degree in technology/industrial arts preferred.
- Employment with a well-known contractor preferred.

That's quite a range of skills, all requiring work, discipline, and study to attain. The job duties are just as varied; the employer asks that the employee perform project installations including on-site design implementation, assembling mounting hardware, mounting equipment, wiring solar systems, and documenting the work done in words and photographs, as well as cleaning up the site, servicing tools and vehicles, and talking about and selling the employer's products "to interested persons as required." That's quite a full plate, and we can only hope that the employer is willing to pay well for the right person to pick it up.

In its listing, the same employer indicates that NABCEP solar PV installer certification will go a long way in meeting all its qualifications. This certification comes from the North American Board of Certified Energy Practitioners (*www.nabcep.org*), a trade association organized, in part, to oversee standards within the field. To acquire this NABCEP certification, a candidate must be at least 18 years of age; meet prerequisites of related experience and/or education; complete an application form documenting requirements; sign a code of ethics; pay applicable fees; and pass a written examination. The NABCEP lays out several paths to meeting the experience and/or education requirement:

- Four years of experience installing PV; OR
- Two years of experience installing PV systems in addition to completion of a board-recognized training program; OR
- Be an existing licensed contractor in good standing in solar or electrical-construction related areas with one year of experience installing PV systems; OR
- Four years of electrical-construction related experience working for a licensed contractor, including one year of experience installing PV systems; OR

Left: A dish Stirling system at a Golden, Colorado test site. Right: A Sandia engineer manually operates the solar concentrator at Ft. Huachuca, Arizona. Photos: Sandia National Laboratory

- Three years experience in a U.S. Dept. of Labor approved electrical-construction trade apprentice program, including one year of experience installing PV systems; OR
- Two-year electrical-construction related, or electrical engineering technology, or renewable energy technology/technician degree from an educational institution plus one year of experience installing PV systems; OR
- Four-year construction related or engineering degree from an educational institution, including one year experience installing PV systems.

Installing PV modules, and then connecting several arrays at a rice drying facility in California. Photo: DC Power / Soligent

Other trades within the solar-energy field involve skill in crafts such as plumbing, electricity, and carpentry. Jobs include the design and maintenance of solar hot-water systems, using the sun to heat domestic hot water as well as buildings themselves; the construction of buildings and communities that use renewable energy; and the retrofitting of older homes and businesses to accommodate new technologies.

Finally, while manufacturing has long been declining in the United States, photovoltaic module assembly is an exception. Even so, most manufacturing in photovoltaics takes place abroad, particularly in China.

. . . .

Reaching Beyond Our Borders with Solar

Delivering Solar Electricity to the Rural Poor

Neville Williams founded the nonprofit Solar Electric Light Fund (SELF) in 1990. It promoted solar power for a decade by setting up pilot solar rural-electrification programs in eleven countries. In 1997 he launched the commercial Solar Electric Light Company (SELCO) in India, Sri Lanka and Vietnam. SELCO-India has brought solar electricity to over 150,000 families. His experiences and insights can be found in his book *Chasing the Sun: Solar Adventures Around the World (www.NevilleWilliams.com)*. Here are a few excerpted passages:

"Ironically, people who never benefited from the age of oil, who never had electricity, are the solar power pioneers who today are using the technology we will all be using in the future. They are building the much-touted 'solar economy'."

"We could deliver an SHS [solar home system] for less than it cost to run an electric line 100 meters from the nearest power main. And this didn't include the cost of transmission or generating the power, just the hookup and house wiring. The sun is, indeed, a better distributor of power than copper wires, especially when houses are widely dispersed, as they are in rural Sri Lanka and in most developing countries. Sri Lanka became our solar learning lab, and today it leads the world in solar installations. One day it may become the world's first solar-powered island."

"Over the years, I personally listened to hundreds of personal testimonials from peasants as they described how electricity changed their lives, how they no longer needed to use kerosene, how much better their children could study at night, and how much they enjoyed seeing television for the first time. One I recall was an old man who said, having seen Jiang Zemin on TV, "I have never before been able to see the emperor!" Another elderly peasant, shedding tears, said, "I have long heard that city folks do not need oil to generate light, but in all my 70 years, this is the first time to actually see such a phenomenon with my own eyes. What a beautiful thing!"

Renewable Energy in the Developing World

Walt Ratterman earned a Master of Science degree in Renewable Energy from Murdoch University in Australia, and was one of the first PV installers to pass the rigorous NABCEP certification exam. He had extensive renewable energy design and installation experience internationally, including Nicaragua, the Galapagos Islands, southern Ecuador, Peru, Arunachal Pradesh in India, Burma, Thailand, and Rwanda. At the time of his tragic death during the Haitian earthquake, he was the Chief Project Officer & Director at SunEnergy Power International (*www.sunepi.org*). The following words he shared with *Home Power* magazine in 2007:

"Here in the United States, solar-electric (that is, photovoltaic) systems typically consist of modules on our roofs, connected to the utility

A solar water pumping project in Pakistan (Solar Energy International and SunEnergy Power International). Photo: Walt Ratterman

grid to generate some portion of our household electricity. But in many parts of the developing world, solar energy is the only source of electricity for a home or a village, because no grid exists.

"In the States, average-sized residential solar-electric systems are between 3 and 5 kilowatts (KW). In the developing world, systems of that size could run an entire village or a large community health center. Average home systems in the developing world are 50 to 75 watts, and 'large' systems may be 120 watts—a fraction of the size of a typical residential system in the United States.

"Working with renewable energy in the developing world is exciting and well worth the rewards. If you have the desire to become involved, keep pushing for answers. The right opportunity will surely present itself. The need remains huge, with more than 1.5 billion people who live with absolutely no electricity. Bringing electricity to these people can make dramatic changes in their lives, allowing them to educate themselves, start businesses, and improve their standard of living.

"Working to give people in the developing world access to electricity is always a two-way street. The people we help have so much to gain by improving their access to education, health care, additional work opportunities, and much more. But we gain as much or more from the experience."

· ·

Knowledge Is Power

It will come as no news to jobseekers, particularly those who are entering the marketplace for the first time or who have been laid off and are seeking to reenter it, that finding suitable employment in this tough economy is not easy. That said, even with the soft employment market between 2008 and 2012, the solar industry added jobs ten times faster than did the overall economy. That was still far from the 26 percent projected for 2012, but the picture was far brighter for workers in solar energy than for those in other fields.

The job market in the renewable-energy sector is expected to increase in strength well into the 21st century, especially for those who enter it with training and certification from an accredited technical school or with formal apprenticeship training. Economists warn that technicians who specialize in installation work may suffer periods of unemployment, since new construction is susceptible to the whims of the market, but that maintenance and repair work will almost always offset times when the economy goes soft. Newer systems tend to be more complex than their predecessors, which is one reason that employers are likely to prefer technical-school graduates or students who have taken appropriate courses at the junior and community-college level.

Fortunately, many secondary and postsecondary technical and trade schools, junior and community colleges, and programs within the armed forces offer training in aspects of solar-power systems, including heating, air-conditioning, and refrigeration. In them, students also acquire at least a basic understanding of electronics. On the trades level, formal apprenticeship programs are also offered through organizations such as the Associated Builders and Contractors (*www.abc.org*), the National Association of Home Builders (*www.nahb.org*), and the International Brotherhood of Electrical Workers (*www.ibew.org*). These programs normally last from three to five years and combine on-the-job training with classroom instruction. Under most circumstances a candidate for them must have a high school diploma or the equivalent and have solid math and reading skills, for in those trades the ability to measure, specify, and communicate clearly are of the utmost importance.

Specialized undergraduate programs in alternative energy, including solar, are few now, but they are likely to become much more common in the near future. One interesting exception is the Photovoltaics and Solar Energy undergraduate program at the University of New South Wales in sun-drenched Australia (see *www.pv.unsw.edu.au*), which offers a five-area approach that it describes as follows:

1. Device and system research and development.
2. Manufacturing, quality control and reliability.
3. PV system design (computer based), modeling, integration, analysis, implementation, fault diagnosis and monitoring.
4. Policy, financing, marketing, management, consulting, training and education.
5. Using the full range of renewable energy technologies including alternate energy technologies (such as wind, biomass and solar thermal), solar architecture, energy efficient building design, and sustainable energy.

The photovoltaic engineering program includes training in technology development, manufacturing, quality control, life-cycle analysis, system design, diagnosis and maintenance, and policy development, all critically important fields. It weds practical and theoretical knowledge and requires industrial training as well as classroom work.

For scientific and theoretical work, undergraduate and, usually, graduate training is expected and even required. At the top end of the schools in the energy field is the Massachusetts Institute of Technology, which counsels that students prepare themselves well in high school for the rigorous work that follows. The MIT web site suggests an academic foundation that includes:

- One year of high school physics
- One year of high school chemistry
- One year of high school biology
- Math, through calculus
- Two years of a foreign language
- Four years of English
- Two years of history and/or social sciences

"Overall, you should try to take the most stimulating courses available to you," the site continues. "If your high school doesn't offer courses that cha llenge you, you may want to explore other options, such as local college extension or summer programs."

If, as an MIT student, you wanted to follow a program in electrical science leading to a specialization in solar-electrical engineering, your coursework would follow a sequence requiring classes and labs in probability and differential equations, as well as electromagnetism, signal processing, and device and circuit design. Graduate specialization might follow several options, depending on the nature of the career you want to follow. The advisors for the University of Colorado graduate program in building-systems engineering put it this way: "Your course of study may be wide or narrow; it will depend on your personal career plans. People who are thinking of a future in research work may take a different approach than those who see themselves working in industry or for a consulting engineering firm."

Like many other graduate programs, Colorado's accommodates students from backgrounds other than engineering, though it recommends undergraduate coursework in calculus, thermodynamics and heat transfer, and building mechanical systems. The program offers concentrations in four areas: energy analysis, HVAC (heating, ventilation and air conditioning) systems, illumination, and solar and renewable energies. The last requires these courses, all listed in the civil engineering department:

- Introduction to Solar Utilization
- Solar Design
- Advanced Solar Design—Photovoltaics
- Advanced Passive Solar Design
- Energy and Environmental Policy
- Analysis and Assessment of Renewable Energy Systems
- Building Systems Seminar

The Colorado catalog counsels, "You should also be (or become) competent in computer programming, word processing and spreadsheet usage to solve engineering problems." Other programs across the country have much the same expectations, so if you're contemplating a future on the R&D end of the alternative-energy spectrum, it's time to hit the books! ❖

Installation of a mirror panel on a concentrating solar power system in New Mexico. Photo: C.E. Andraka, Sandia National Lab

Resources

Organizations & Programs

American Solar Energy Society

www.ases.org

(303) 443-3130

Founded in 1954, ASES boasts more than 10,000 members across the nation who are at work creating a sustainable energy economy. Their web site also serves as a meeting place for qualified professionals who are able to design, install, and service solar-energy systems. ASES publishes Solar Today magazine, a flagship publication in the field. Job listings are also posted.

Findsolar.com

www.findsolar.com

A web site that will connect you to renewable energy professionals; sponsored by ASES, Solar Electric Power Association, and Department of Energy.

Florida Solar Energy Center

www.fsec.ucf.edu

Established in 1975, FSEC is the largest and most active state-supported renewable energy research institute in the United States. It is an accredited laboratory for testing and certification of solar technologies.

International Solar Energy Society

www.ises.org

ISES has been serving the renewable energy community since 1954. A UN-accredited NGO present in more than 50 countries, the Society supports its members in the advancement of renewable energy technology, implementation and education all over the world.

Midwest Renewable Energy Association

www.midwestrenew.org

(715) 592-6595

MREA promotes renewable energy, energy efficiency, and sustainable living through education and demonstration. It offers workshops in renewable energy, including wind systems; photovoltaics; solar thermal systems for domestic hot water and space heating; plus photovoltaic, wind, and solar thermal site assessor training and certification.

National Renewable Energy Laboratory, Solar Research

www.nrel.gov/solar

The resources that follow are given in alphabetical order. We make an exception for this first listing, the NREL website devoted to solar research. Use this as your starting point, for no matter what your interest in studying and working in the field, NREL will be of help in providing well-researched, timely information.

North American Board of Certified Energy Practitioners (NABCEP)

www.nabcep.org

(800) 654-0021

As its site says, "NABCEP is committed to providing a certification program of quality and integrity for the professionals and consumer/public it is designed to serve. Professionals who choose to become certified demonstrate their competence in the field and their commitment to upholding high standards of ethical and professional practice."

Power from the Sun

www.powerfromthesun.net

This site offers an online, revised and updated, and free version of Solar Energy Systems Design, a highly useful textbook first published by W. B. Stine and R. W. Harrigan in 1985. It contains other resources that are useful for learning about solar energy, including problem sets and various calculators.

Solar Energy International
www.solarenergy.org
(970) 963-8855
SEI is a non-profit organization offering accredited online and hands-on workshops in solar, wind, and water power and natural building technologies in locations around the world. SEI provides the expertise to plan, engineer, and implement sustainable development projects, and it offers a jobs-listing board for SEI alumni.

Solar Living Institute
www.solarliving.org
(707) 472-2450
Established in 1998, the Solar Living Institute is a nonprofit educational organization whose mission is to promote sustainable living through inspirational environmental education. The Institute provides practical education by example and a range of hands-on workshops on renewable energy, green building, construction methods, and other subjects.

Publications

Solar Electricity Handbook by Michael Boxwell, 2012 edition (Greenstream Publishing, 2012). Writes a satisfied reader at Amazon.com, "This is a well-written book with exactly the details you need if you are planning your own DIY solar project, or preparing to buy or contract out for a solar power system. I bought several books on the subject, and this is the only one to answer my very specific questions, such as where to put fuses and isolation switches. It seems up to date and gives the names of respected manufacturers."

Photovoltaic Systems, 2nd edition by James P. Dunlop, (American Technical Publishers, 2009). A comprehensive textbook for the design, installation, and evaluation of residential and commercial photovoltaic (PV) systems, both stand-alone and grid-tied systems. The included CD-ROM features interactive resources for independent study and to enhance learning.

Got Sun? Go Solar by Rex A. Ewing and Doug Pratt, (PixyJack Press, 2005). Meant for homeowners and small-systems designers, this user-friendly handbook of grid-tie solar gives a very good idea of the kinds of skills and training that go into making an energy-efficient building.

Power Surge: Guide to the Coming Energy Revolution by Christopher Flavin and Nicholas Lenssen, (Norton, 1994). An interesting discussion of the technologies discussed throughout this book, Flavin and Lenssen's overview suffers only from being dated. Anyone who aspires to a career in alternative energy, however, ought to give it a careful reading.

Home Power Magazine is dedicated to renewable energy and sustainable living solutions. This magazine provides extensive product information, homeowner testimonials, buyer advice, and "how-to" instructions. *www.homepower.com*

Passive Solar House: The Complete Guide to Heating and Cooling Your Home by James Kachadorian, (Chelsea Green Publishing Company, 2006). Applicable to diverse regions, climates, budgets, and styles of architecture, this book offers proven techniques for building homes that heat and cool themselves. Includes CSOL passive solar design software.

Solar Water Heating: A Comprehensive Guide to Solar Water and Space Heating Systems by Rob Ramlow and Benjamin Nusz, (New Society Publishers, 2006). This books presents the basics of solar water heating, including solar water and space heating systems, system components, installation, operation, maintenance, and system sizing and siting.

The Solar Economy: Renewable Energy for a Sustainable Global Future by Hermann Scheer, (Earthscan Publications Ltd, 2004). A powerful and comprehensive account of how the global economy can and must replace its dependence on fossil fuels with solar and renewable energy—and the enormous and multiple benefits that should follow.

Photovoltaics: Design and Installation Manual by Solar Energy International, (New Society Publishers, 2004). This sturdy reference work covers the basics of solar electricity, photovoltaic applications and system components, solar site analysis and mounting, component specification, and safety issues, among many other topics.

Chasing the Sun: Solar Adventures Around the World by Neville Williams, (New Society Publishers, 2005). A fascinating account of the author's twelve-year quest to bring solar power and light to people in the developing world who have no electricity. Illuminating reading to all interested in the environment, development, renewable energy, socially responsible business, and our future at the end of the age of oil.

Solar Energy International's "Women and Renewable Energy Program" provides an opportunity to break the chain of gender bias. Training programs teach renewable energy technologies to equip women with the necessary skills to participate in the design, installation and maintenance of systems. "We've had NASA engineers, licensed electricians, schoolteachers, and homemakers attend our women's PV Design & Installation workshops," says Laurie Stone of SEI. Photos: Solar Energy International

Where To Study

California Institute of Technology

www.caltech.edu
1200 East California Blvd.
Pasadena, CA 91125
(626) 395-6811

Caltech is a leading research institution in the study of renewable energy, especially solar energy and biofuels. Its chemistry department, to name just one program, sponsors research in solar energy conversion and storage, methane oxidation, and nitrogen fixation.

Coconino Community College

www.coconino.edu
Lone Tree Campus
2800 S. Lone Tree Road
Flagstaff, AZ 86001–2701
(928) 527-1222, (800) 350-7122

CCC offers a certification program as an alternative energy technician, with an intermediate certificate requiring 28–29 credit hours and an advanced certificate requiring 52–57 credit hours.

Colorado State University

www.warnercnr.colostate.edu
Warner College of Natural Resources
103 Natural Resources Building
Fort Collins, CO 80523
(970) 491-5629

Warner College of Natural Resources is committed to offering a comprehensive range of undergraduate and graduate degree programs that directly address today's most important environmental and natural resource issues. Its programs are grounded in state-of-the-art science and technologies and involve students in direct problem-solving experiences. Its students are well prepared to become leaders in environmental and natural resources management and science. According to the Association for the Advancement of Sustainability in Higher Education (AASHE), CSU is one of the top four public research universities for solar power, as well as sixth in the production of solar power on a home campus.

Illinois Institute of Technology

www.iit.edu
Energy/Environment/Economics
3300 South Federal Street
Chicago, IL 60616-3793
(312) 567-3000

E3 is an academic program of research and coursework for students in chemical, mechanical, environmental, and electrical engineering. The research program encompasses areas of specialization that relate to energy, sustainable development, industrial ecology, and environmental design.

North Carolina State University

ncsc.ncsu.edu
North Carolina Solar Center
1575 Varsity Drive
North Carolina State University
Raleigh, NC 27606
(919) 515-3480

Created in 1988, the North Carolina Solar Center serves as a clearinghouse for solar and other renewable energy programs, information, research, technical assistance, and training for the citizens of North Carolina and beyond. Students at NCSU have opportunities to learn at the center, and the school also offers excellent programs in many other renewable energy technologies.

Oklahoma City Community College

www.okc.cc.ok.us
7777 South May Avenue
Oklahoma City, OK 73159-4444
(405) 682-1611

OKCC's program prepares individuals to apply basic engineering principles and technical skills in support of engineers and other professionals engaged in developing solar-powered energy systems. Includes instruction in solar energy principles, energy storage and transfer technologies, testing / inspection procedures, system maintenance procedures, and report preparation.

Pennsylvania College of Technology

www.pct.edu
One College Avenue
Williamsport, PA 17701
(570) 326-3761

PCT offers two- and four-year courses in solar technology, electric-power-generation technology, electrical technology, and other areas relevant to renewable energy.

San Juan College

www.sjc.cc.nm.us
4601 College Blvd.
Farmington, NM 87402
(505) 327-5705

The Renewable Energy Program gives the student a solid foundation in the fundamental design/installation techniques required to work with renewable technologies. The concentration in Passive Solar Design and Analysis is offered as an A.A.S. degree and/or a one-year certificate. Both avenues include hands-on electrical training in a modern computer-based laboratory setting where the National Electrical Code (NEC) is emphasized throughout the curriculum.

University of Central Florida

www.fsec.ucf.edu
Florida Solar Energy Center
1679 Clearlake Road
Cocoa, FL 32922–5703
(321) 638-1000

FSEC is the largest and most active state-supported renewable energy research institute in the United States. It is an accredited laboratory for testing

An NREL engineer tests the output of PV panels. Photo: Holly Thomas, NREL

and certification of solar technologies. Established in 1975, it also conducts research in building science, solar energy, hydrogen and alternative fuels, fuel cells, and other advanced energy technologies. The school offers courses and workshops that take advantage of this renowned resource.

University of Colorado Environmental Center

ecenter.colorado.edu
University Memorial Center, Rm 355
Boulder, CO 80309-0207
(303) 492-8308

Colorado is one of the greenest campuses in the nation, literally and figuratively, and a leader in the use of renewable energy sources such as wind and solar power. The university offers a wide range of programs in every aspect of renewable energy.

University of New South Wales

www.pv.unsw.edu.au
School of Photovoltaic and Renewable Energy Engineering
Electrical Engineering Building
Sydney NSW 2052
Australia
+61 2 9385 1000

The School of Photovoltaic and Renewable Energy Engineering delivers the world's first photovoltaic and renewable energy engineering degree programs. The School includes the Australian Research Council Centre for Advanced Silicon Photovoltaics and Photonics (also known as the ARC Photovoltaics Centre of Excellence).

University of Wisconsin

sel.me.wisc.edu
Solar Energy Program
1343 Engineering Research Bldg.
1500 Engineering Drive
Madison, WI 53706-1687
(608) 263-5626

For study at the Solar Energy Laboratory, which offers MS and PhD degrees, applicants must have a bachelor's degree in engineering or its physical science equivalent, have a strong undergraduate record and be admitted to either the mechanical or chemical engineering department. Three letters of recommendation are required. The GRE examination is recommended.

Wind Energy

Since the early 1990s, wind power has become one of the fastest-growing sources of electricity generation in the United States, while other nations have enthusiastically pursued wind power as well. As of this writing, wind turbines have been put in place in 39 states, both onshore and offshore, with more joining the list each year. California and Texas lead the country, but places such as Ohio and New York are making significant advances in adding wind-generated energy to the overall energy mix.

Photo: Vestas Wind Systems A/S

The US Department of Energy estimates, in fact, that wind-generating capacity in the United States, which is now second only to China's, may supply 20 percent of US demand by 2020. Given that as of 2012 wind power met only 3.6 percent of domestic demand, to accommodate that expansion suggests that the sector will need to expand markedly in the next few years—and that in turn means jobs.

Wind energy, investors and homeowners alike are finding, is clean, efficient, and inexhaustible. In addition, a single commercial wind turbine of ordinary size—about 250 feet across, that is—can generate some 2 megawatts at peak output; a solar array of the same size can generate only a fraction of that power and, at present, would cost several times more to put on line. It is for those qualities that states, municipalities, and private utilities are flocking to wind power.

According to the American Wind Energy Association (AWEA), jobs in wind energy fall into several areas, such as manufacturing and engineering, environmental and consulting services, and sales and marketing. In addition, research and academic positions are likely to grow in number. The situation, in other words, is very much like that of the solar-energy field, and major players in the energy industry, such as General Electric, Siemens, and Shell Oil, are now heavily involved in research.

All of this suggests a bright future for wind energy, and for those who work in it. For the near future, most new wind-energy jobs will be

Installing turbines by crane at the Horns Rev Offshore Wind Farm in the North Sea.
Photo: Elsam A/S (www.hornsrev.dk)

in manufacturing, installation, and operations: in building, siting, setting up, and maintaining and repairing wind turbines and associated equipment. Some manufacturing will involve producing the giant wind towers that you see scattered across the landscape, but there is also growth projected in home-energy systems, small towers only a quarter or a third of the height of the giants, but capable of supplying much of a household's demand for electricity. There are also opportunities in areas such as business development, marketing, sales, computer modeling, and consulting, and plenty of places to train for careers in the field and enhance one's set of skills.

The following table, set in a typical wind turbine manufacturing company in the United States, suggests the range of jobs available. The salary information dates to 2006, but given generally stagnant wage and salary growth since 2007, it remains current. As you will see, there is a sharp distinction between skilled and unskilled jobs and between jobs requiring only secondary or technical education and those requiring a degree.

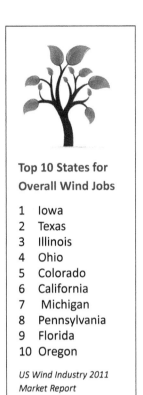

Top 10 States for Overall Wind Jobs

1 Iowa
2 Texas
3 Illinois
4 Ohio
5 Colorado
6 California
7 Michigan
8 Pennsylvania
9 Florida
10 Oregon

US Wind Industry 2011 Market Report

Typical Employee Profile of a 250-Person Wind Turbine Manufacturing Company, 2006

Occupation	Employees	Annual Earnings
Engine and other machine assemblers	31	$36,300
Machinists	27	$40,500
Team assemblers	16	$30,100
Computer-controlled machine tool operators	12	$40,600
Mechanical engineers	10	$71,600
First-line supervisors / Managers of production / operating	10	$59,600
Inspectors, testers, sorters, samplers, and weighers	8	$40,400

Lathe and turning machine tool setters, operators, tenders	6	$40,000
Welders, cutters, solderers, and brazers	4	$39,800
Laborers and freight, stock, and material movers	4	$29,800
Maintenance and repair workers	4	$44,100
Tool and die makers	4	$43,600
Grinding / lapping / polishing / buffing machine tool operators	4	$34,800
Multiple-machine tool setters / operators / tenders	4	$40,800
Industrial engineers	3	$70,400
Engineering managers	3	$108,300
Shipping, receiving, and traffic clerks	3	$32,100
General and operations managers	3	$120,600
Industrial production managers	3	$93,100
Industrial truck and tractor operators	3	$34,200
Purchasing agents	3	$56,200
Cutting / punching / press machine setters / operators / tenders	3	$31,400
Production, planning, and expediting clerks	3	$45,200
Milling and planing machine setters / operators / tenders	3	$40,600
Mechanical drafters	2	$39,900
Customer service representatives	2	$39,100
Bookkeeping, accounting, and auditing clerks	2	$35,600
Office clerks, general	2	$29,400
Sales representatives, wholesale and manufacturing	2	$55,300
Janitors and cleaners	2	$29,800
Accountants and auditors	2	$59,800
Sales engineers	2	$72,500
Tool grinders, filers, and sharpeners	2	$44,000
Executive secretaries and administrative assistants	2	$43,200
Mechanical engineering technicians	2	$50,900
Electricians	2	$49,600
Other employees	48	$49,700
Total employees (126 occupations)	**250**	**$46,400**

Source: American Solar Energy Society, 2007

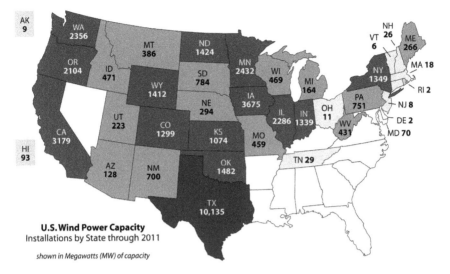

U.S. Wind Power Capacity
Installations by State through 2011

shown in Megawatts (MW) of capacity

As of July 2012, over 51,630 megawatts (MW) of wind power capacity has been installed in the United States. Source: www.awea.org

Careers in Wind Energy

As a comparatively new field, wind energy has not yet solidified into easily defined career tracks. As with solar energy, a technician is often expected to know something of the big picture, with an understanding of trends in the business and promising research. A salesperson, similarly, needs at least a basic understanding of the technologies involved and of the challenges and benefits associated with, say, putting a wind turbine in place. A technical writer will need to know just about everything in a wind-power business, almost as much as a CEO or CFO. And an engineer working in wind power needs to have practical grounding in how a system that looks good on paper can be put up and maintained without bringing the people who do that work to despair. In short, a successful career in wind energy is going to involve a little bit of everything. The person with wide-ranging interests and abilities is going to thrive in this new green-collar world, while the person who can't be bothered to acquire new skills and keep up with developments in the field is likely to hit a dead end very soon.

Administrative, Financial, and Nontechnical

As in most aspects of the energy industry, wind power relies on engineers and technicians, but also on people with business, financial, and commu-nications skills. Holders of bachelor's and master's degrees in business administration, for example, are in demand, while industry insiders also observe that liberal-arts majors with some aptitude for science, business, and math flourish in the industry. Technical writers are also in demand, as well as writers who can explain wind energy and develop campaigns to promote its use.

The American Wind Energy Association (AWEA), for instance, recently advertised for a junior position in advocacy writing, a job that calls for the ability to match industry experts and journalists needing background, as well as writing opinion articles and letters to the editor. The job calls for "strong oral and exceptional written communication skills; detail ori-ented; ability to proofread; ability to juggle multiple/overlapping projects and meet deadlines; ability to work well with teams; ability to self-moti-vate; ability to be flexible; ability to establish a clear focus on results." It also requires a bachelor's degree in journalism, communications, English, or marketing, with a preference for applicants with a minimum of three years of communications, marketing, or advocacy work within the wind industry or the nonprofit sector.

Jobs in Wind Energy

There are 75,000 jobs in the wind industry, existing across all 50 states. Over 400 manu-facturing facilities in the US supply some of the 8,000 components in a wind turbine.

Source: US Wind Industry 2011 Market Report

Research

At present, according to the AWEA, research opportunities in wind energy center on five broad areas:

- **Turbine research:** research to improve turbine design (aerodynam-ics), understanding the nature of wind (inflow and turbulence), and using computer models to design efficient and low-cost turbines (mod-eling structures and dynamics).

- **Wind resource assessment:** mapping an area to include specific wind data such as average wind speed and variability.
- **Forecasting**: using weather models and technologies to predict wind speeds and patterns at various altitudes.
- **Utility grid integration:** research to improve the ways in which energy produced by wind is introduced into the utility grid.
- **Energy storage:** developing technology to store wind energy as electricity.

Resource assessment is particularly important and provides many opportunities. It does no good to erect a wind turbine in an area in which the wind does not blow, and trained assessors are essential to the work of determining where to locate wind-generating stations and predicting how much power they will contribute to the mix. Often, trained meteorologists shift over to such work, which combines theoretical knowledge and field experience: another green-collar blend, in other words. Other desired backgrounds are computer science, aerodynamics, physics, and mathematics.

Because of federal and state environmental-impact requirements, new sites must also be studied to determine how wind turbines might affect water supplies and plant and animal communities. Wind towers are known, for example, to produce high rates of mortality in migratory bird

Top 10 States for Wind Projects under Construction in 2012

Kansas (1,189 MW)

Texas (857 MW)

California (847 MW)

Oregon (640 MW)

Illinois (615 MW)

Pennsylvania (520 MW)

Iowa (470 MW)

Oklahoma (393 MW)

Michigan (348 MW)

Washington (331 MW)

species, and utilities are actively seeking to site towers in ways to reduce this problem. This provides research and field opportunities for workers with bachelor's or graduate degrees in biology or environmental science, though most scientific jobs demand a master's degree or higher.

Here are two listings for jobs within the broad field of assessment. As you'll see, both call for wide-ranging combinations of skills, experience, and education. The first is from Airtricity, a leading renewable-energy utility that is active throughout North American and the British Isles:

This position reports to the WARS (Wind Analysis, Research & Standards) U.S. Manager and will be an active member of the WARS team: Early resource assessments and layouts; Occasional site visits to ascertain terrain, roughness and obstructions; Assistance with quality control and cleaning of large data sets from over 70 meteorological masts on a regular (3x weekly) basis; Assistance with delivery of high quality datasets of wind measurements from each site suitable for use in a low uncertainty wind resource assessment; Internal electricity production estimates using WAsP & WindFarmer, final layouts and management of subcontracted analyses; Internal site condition assessments; Use of GIS tools such as ArcView, Delorme XMap and GlobalMapper to produce high quality maps; Preparation of site reports; Implementing WARS safety standards for meteorological masts; Possibly overseeing power performance testing of wind turbines. Position Requirements: BS in Physical Science, Atmospheric Science, or Mechanical Engineering; Preferably 1–3 years post-qualification experience in the wind resource assessment field.

The second, at a higher level, is from Black & Veatch, a design and manufacturing firm. It seeks an experienced engineer to assist with and manage all aspects of wind energy project assessment and development. Key responsibilities of this position include:

- Develop, plan, coordinate and manage all aspects of wind and other renewable energy projects.
- See projects from concept through completion. Areas of emphasis may include: Utility-scale wind farms, community wind systems, distributed wind, and offshore wind projects.

- A wide variety of diverse assignments including resource assessment, due diligence, siting, feasibility studies, new technology evaluation, project proposal evaluation, conceptual design, design review, performance estimates, and cost estimates.
- Interaction with clients including utilities, energy project developers, financial institutions, energy end-users, investors, government agencies, and other organizations.
- Manage projects, including scope definition, project staffing, technical oversight, and responsibility for successful project execution on time and within budget.
- Lead / contribute to authorship of reports, studies, and proposals.
- Manage work of others in multi-disciplinary teams.
- Lead and assist with business development activities in area of expertise. Assist with development of new project opportunities and business lines.
- Exposure to wind resource analysis and production estimation. Although the primary focus of the job will be wind energy projects, the position would include exposure to all renewable technologies: biomass, wind, geothermal, solar, hydro, etc.

The preferred candidate will possess:
- Engineering degree from an ABET-accredited university. Masters degree or MBA would be beneficial. Some course work on energy systems, specifically some knowledge of renewable energy technologies preferred.
- Minimum 8 years experience with wind energy projects.
- Excellent verbal and written communication skills.
- Ability to communicate clearly and succinctly through formal reports, presentations, memoranda and email.
- Must be able to function equally well in collaborative, multi-discipline teams, and self-directed independent assignments.
- Must be self-motivated, with an ability to balance multiple projects while working under tight deadlines.
- Experience in wind energy projects specifically.
- An interest in all renewable energy technologies.
- Experience with conceptual and detailed design is desired. Project field experience is also a plus.

Here is a posting from June 2012, from Britain, for an offshore environmental assessment post. In it, EIA stands for environmental impact assessment; it is formally equivalent to the EIS, or environmental impact statement, in the United States:

- Position Overview: Reporting to the Senior Environmental Planner the successful candidate will be responsible for supporting the development of all onshore aspects related to offshore wind farm development.
- Responsibilities: Site Selection; Perform EIA studies; Consent applications; Development of cable landfalls; Ensure the project is completed on time, in budget with minimum impact to the environment and stakeholders.
- Education: Educated to degree level preferably in an environmental related subject.
- Core Requirements: Minimum of 5 years experience with EIA's; Onshore environmental planning, minimum 3 years; Knowledge of planning application preparations related to renewable energy infrastructure projects; You must hold a valid driving license.
- General Requirements: Membership of IEMA or IEEM or equivalent would be advantageous but not essential; Well rounded knowledge of onshore environmental issues; Accuracy in the delivery of documentation and works; Good communication skills.
- Key Skills: Wind, EIA, Onshore Development, Project Planning, Project Development, Offshore Wind.

All three positions, it is plain to see, require a solid educational background and significant experience in the field. Getting there is a challenge, but there are plenty of entry-level positions in every aspect of wind power that will set you on the right road.

Technical

The sector in wind energy that will produce the majority of new jobs in the next decade is manufacturing, installation, operation and maintenance; the demand for new plants is great, so much so that new clients often are put on a waiting list for the next available units. The manufacturing sector tends to hire trained engineers with degrees from four-year colleges, though two-year programs are increasingly contributing to the labor pool.

A two-year background is sufficient, for instance, for workers who produce blades, towers, and gearboxes, the stuff of which wind towers are made; the control systems for them are electrical, therefore requiring training in electronics. Here, for example, is a job listing for a position as a "commissioning technician":

An international renewable energy equipment provider is seeking several Wind Turbine Commissioning Technicians for direct hire positions. Positions are available in PA, TX, & IL. Technician troubleshoots mechanical and electrical problems on variable-pitch, variable-speed turbines. Assists in all areas of site construction and commissioning as directed by the designated site supervision. Performs all mechanical and electrical component testing and ultimately repairs as needed. The position requires knowledge of direct current electricity / alternating current electricity. The technician must be able to consistently climb wind turbine towers, to consistently travel for up to six weeks at a time between short trips home equaling two days and to work outdoors under extreme weather conditions. Extensive travel and overtime is required when needed to meet project schedules.

The wind-power industry also offers ample opportunities for other field technicians, installation technicians, and maintenance workers. Here is an advertisement from Britain calling for a wind tower and turbine inspector:

- In-service inspection and periodic monitoring
- End of warranty inspections
- Commissioning inspections

Left: Service work on a commercial turbine. Photo: Vestas Wind Systems A/S
Right: Preparing the tower base. Photo: NEG-Micon

- Inspection for condition based maintenance
- Blade inspections
- Vibration measurements and endoscopic gearbox inspections
- Maintenance management
- Construction supervision
- Damage investigation
- Communicate with the experts internally and externally
- Establish and maintain a project file and documentation system
- Assuring required quality of inspections
- Establish long term relationship with the clients
- Prepare quotations, scopes of work and check lists
- Assure inspections are carried out according to the requirements
- Prepare and compile inspection reports
- Relevant experience preferable in more than one of these fields:
  - Wind Turbine Technology and Engineering
  - Gearbox and rotor blade technology
  - Condition monitoring systems (CMS)
  - Vibration measuring techniques
  - Endoscope inspection experience
  - Electrical installations (generator, power converter, transformer, cabling, etc.)
- Ability to work at heights
- Valid safety training (incl. HSE) for wind turbines
- Ability or valid training/ certification for application of rope access technique for rotor blade inspection
- Able to cooperate effectively with multidisciplinary and multinational teams
- Good communication skills
- Initiative and ability to work proactively and independently
- Flexibility and efficient working attitude
- Team player
- Educational background as Engineer, in Mechanical/Electrical Engineering and with 3-5 years relevant experience within the wind industry. Ideally with specific experience as a wind turbine inspector with onshore and offshore experience.

"Ability to work at heights" is somewhat of an understatement, given that blade inspections require the inspector to rappel down the blade and from there nearly to the ground!

All these jobs, many set at residential as well as commercial sites, have different requirements. Some call for one- or two-year certificates, others bachelor's degrees. At the most basic level of manufacturing and installation, a high-school diploma is sufficient, though these basic jobs tend to pay far less than other positions—often no more than $15.00 an hour to start. Such positions have requirements of their own, of course, including, as the education ministry of the wind-rich Canadian province of Alberta puts it, "Skills in math, reading, and writing in order to read, interpret, and work from complex instructions, diagrams, and prints; computer literacy; capable of using and operating various equipment including torque wrenches and basic electrical test equipment (electrical or mechanical training is preferred); fiberglass repair skills; good physical condition (some physical activities include ladder climbing and heavy lifting); the ability to work at substantial heights (up to and exceeding 185 feet); problem solving, decision-making, and troubleshooting abilities; ability to work as part of a team as well as independently; must be self-motivated, energetic, and possess a positive attitude; a commitment to safe work habits; safety certification."

Small Wind Power

Most of us are familiar by now with the concept, if not the sight, of 300-foot-tall wind towers that can power whole city blocks. Small wind turbines, generally providing less than 100 kilowatts, are less common, but, as the Small Wind Certification Council (www.smallwindcertification.org) notes, they "have great potential to serve homes, farms, schools, and other end-users." Depending on local zoning ordinances and codes, these small wind towers can reach heights of 75 feet, and occasionally even taller, providing electrical power, generally, to single buildings or farms. A friend of mine, an Arizona rancher, powers a whole series of residential and livestock-related structures with a combination of two small wind towers and solar panels; that pairing keeps half a dozen humans and more that three hundred animals safe and comfortable.

"Small wind" training programs are somewhat less common than programs for their larger, industrial-scale counterparts, but there are still numerous good ones across the country. For an idea of curriculum, visit Appalachian State University's North Carolina Wind Energy site. For other training possibilities, consult the references available at *Windustry.org*. And for an overview of the combined educational and technical requirements involved, see the site of the chief certifying organization, the North American Board of Certifying Energy Practitioners. The NABCEP Candidate Information Handbook may seem daunting, but it provides an excellent blueprint for planning a curriculum and timetable for self-study, if a formal program is not available in your area.

Installers of small wind systems must have a broad-based knowledge of site assessment for wind conditions, how to install and maintain small-wind

turbines and towers, and other technical areas, as well as the "people skills" necessary to sell wind systems and keep their customers happy. Since these small systems are often located in rural areas and on farms scattered hither and yon, it helps to be willing to travel—and travel on bumpy dirt roads at that. Small wind systems are also often combined with solar electricity, so a knowledge of solar power systems is a plus.

Appalachian State University's North Carolina Wind Energy
wind.appstate.edu/education-training
Great Plains Windustry Project
www.windustry.org/where-can-i-find-a-school-or-training-program-specific-to-renewable-energy
North American Board of Certifying Energy Practitioners
www.nabcep.org/certification/small-wind
Use their search engine to find the latest candidate information handbook.

Knowledge Is Power

Many positions in wind energy research require an engineering background and a four-year degree in some aspect of engineering. For example, Amir Mikhail, the chairman of wind energy company Zond, holds a bachelor's and master's degree in aerospace engineering, the latter from Georgia Institute of Technology; his early research, extrapolated from his knowledge of air flow over airplane wings, was on wind speeds around the world. Someone with an engineering background might, for instance, soon solve one of the most vexing problems in wind energy: how to store the electricity thus generated. (Currently proposed solutions include compressed air and advanced batteries.) Renewable energy problems, like so much else in the economy, require innovative thinking; as Siemens USA CEO Eric Spiegel remarks, "Constant innovation is the only way to stay ahead of competitors."

Other positions, as we have seen, involve environmental sciences, meteorology, physics, mathematics, and business. The most capable players in the field, those who are likeliest to succeed in the coming years as the business grows and changes, will combine technical abilities, theoretical knowledge, people skills, business smarts, and a hunger to know all there is to know about renewable energy. In arriving there, an ideal degree might then involve engineering and other sciences, business, communications, and the liberal arts.

Several degree-granting programs offer training in wind energy—Texas Tech University Wind Science and Engineering Research Center, the University of Massachusetts Center for Energy Efficiency and Renewable Energy, the MIT Laboratory for Energy and the Environment, Oregon State University, and the Illinois Institute of Technology Wanger Institute for Sustainable Energy Research among them—and many more offer laboratory programs that allow for guided independent study in the area. Some programs, such as that offered by Iowa State University, are intensive summer courses—in the case of ISU, a ten-week program in wind energy science, engineering, and policy. Georgia Tech's Strategic Energy Research program allows for specialization in several renewable energy fields, with thirteen faculty members and researchers devoted to wind energy alone.

Most of these programs have a small number of core courses followed by intensive study in one's chosen area of specialization. At Texas Tech's Wind Science and Engineering Research Center, for example, a degree involves just five core courses (with the home departments in parentheses):

- Wind Science (atmospheric science)
- Wind Engineering (engineering)
- Economic Policy (economics)
- Statistics for Engineers I and II (engineering)
- Leadership and Ethics (engineering)

These core courses are followed by specializations in electrical engineering, civil engineering, mechanical engineering, atmospheric sciences, geology, management, economics, and other fields.

Several community colleges offer programs leading to an associate degree or certification. These programs are not yet as widely developed as those for solar energy, but their numbers are growing. One particularly interesting program, and a model for others, is offered by the Iowa Lakes Community College. The ILCC Program in Wind Energy and Turbine Technology graduates about thirty students a term, and almost all of them enter the industry immediately. Taking the first three terms, for 48 units, leads to a diploma; completing the whole course of 81 units leads to an associate's degree in applied science.

ILCC Program: Wind Energy and Turbine Technology

First Term
Field Training and Project Operations
Direct Current Electrical Theory
Alternating Current Electrical Theory part I
Introduction to Wind Energy
Introduction to Computers/Info Systems
Intermediate Algebra

Second Term
Human Relations
Alternating Current Electrical Theory part II
Wind Turbine Mechanical Systems
Basic Hydraulics
Business Communications
Electric Motors and Motor Controls Fundamentals

Third Term
Wind Turbine Site Construction / Locations
Wind Turbine Internship

Fourth Term
Power Generation and Distribution
Basic Networking and Computer Technology
Meteorology (or Chemistry or Physics)
Electrical Practical Applications
~~Airfoils & Composite Repair~~

Fifth Term
Data Communication and Acquisition
Programmable Logic Control Systems
Wind Turbine Siting
Principles of Management

Riverland Community College, with three campuses in southern Minnesota, offers a similar program leading to a diploma as a wind turbine technician. It describes the program so:

The Wind Turbine Technician diploma program will offer students a new career path within the Riverland Construction Electrician and Industrial Maintenance and Mechanics Programs. This new option will give students a choice of training to be an entry-level Construction Electrician, Industrial Maintenance Mechanic, Maintenance Electrician and/or now as a Wind Turbine Technician. Wind Turbine Technicians play a key role in ensuring quality, safety and service involving the operation and maintenance of wind turbine units, performing mechanical and electrical troubleshooting, as well as repair and preventative maintenance. Work may include basic circuits, electrical motors and their controls, electronic controls, programmable logic controllers and variable frequency drives. Wind Turbine Technicians install and maintain, repair and replace malfunctioning parts and equipment, transmissions and drives, programmable logic controllers and blade maintenance and repair. A solid background in technical math, computer literacy, problem solving, decision making and troubleshooting abilities, ability to work as a part of a team as well as independently and good reading comprehension are important skills for applicants to this program.

Apprenticeship and internship is another good avenue into a career in wind energy. Visit the Wind Energy Career Center at the AWEA web site (see the resources below) for a list of available internships and job openings. At all levels of education and experience, it's a good idea for professionals to give themselves continuing education through workshops, seminars, and conferences, also listed on the AWEA site. ❖

Wind Energy Quick Facts

A typical turbine today provides 15 times more electricity than a typical turbine in 1990, and can supply the electricity for the equivalent of 500 American homes.

The US wind fleet provides enough electricity to supply 10 million homes, or the same amount of electricity as 10 nuclear power plants.

Source: US Wind Industry 2011 Market Report

Resources

Organizations & Web Sites

American Wind Energy Association
www.awea.org
(202) 383-2500
AWEA, a trade and advocacy organization based in Washington, D.C., represents the U.S. wind energy industry, sponsoring an annual national conference devoted solely to wind energy, which features technical papers from experts in all areas of the field. Visit their Job Board for a wealth of information.

European Wind Energy Association
www.ewea.org
EWEA is the voice of the wind industry in Europe and worldwide. Its members, from 40 countries, include over 300 companies, associations, and research institutions. These members include manufacturers covering almost all of the world wind-power market, component suppliers, research institutes, contractors, national wind and renewables associations, developers, electricity providers, finance and insurance companies and consultants. The EWEA web site is full of useful information on every aspect of wind power.

Great Plains Windustry Project
www.windustry.org
(612) 200-0331 or (888) 818-0936
GPWP offers an information-rich site with advocacy for community wind projects, which are locally owned by farmers, investors, businesses, schools, utilities, or other public or private entities. The key feature, they note, is that local community members have a significant, direct financial stake in the project beyond land-lease payments and tax revenue.

U.S. Department of Energy Wind Power Program
www1.eere.energy.gov/wind/
This program is leading the nation's efforts to improve wind energy technology so that it can generate competitive electricity in areas with lower wind resources, and to develop new, cost-effective, advanced power technologies that will have enhanced environmental performance and greater energy efficiencies.

National Renewable Energy Laboratory
www.nrel.gov
A key unit for energy research within the U.S. Department of Energy, NREL maintains a high-quality, information-rich web site. See, for instance, the page "Wind Energy Basics: How Wind Turbines Work," a concise introduction to a complex subject.

Publications

Wind Energy Handbook by Tony Burton, Nick Kenkens, David Sharp, and Ervin Bossanyi (Wiley, 2011). Perhaps the one book to have for wind energy studies: a reference book, textbook, and survey of current knowledge in the field.

Wind Energy Basics, 2nd ed. by Paul Gipe (Chelsea Green, 2009). Includes chapters on how to evaluate modern wind turbine technology, installing wind turbines safely, designing a stand-alone power system for living off-the-grid, and how to use electricity-producing wind turbines to pump water in rural areas.

Wind Energy Explained, 2nd ed. by J. F. Manwell, J. G. McGowan, and A. L. Rogers (Wiley, 2010). A comprehensive survey of wind energy, presupposing a solid basic understanding of mathematics and physics.

Wind Energy: Fundamentals, Resource Analysis and Economics by Sathyajith Mathew (Springer, 2006). A thorough analysis of wind-energy economics, with an emphasis on material of great use to "green-collar" workers whose work blends technical and nontechnical subjects.

Wind Power Monthly is a source of independent international news about developments in wind and other renewable energies. Subscribers receive access to industry statistics not easily found elsewhere. *www.windpower-monthly.com*

Wind turbines at the KILI radio station, Pine Ridge Reservation. Photo: Bob Gough

Where To Study

Georgia Institute of Technology
www.gatech.edu
North Avenue
Atlanta, GA 30332
(404) 894-2000

Georgia Institute of Technology is a national leader in energy research, offering undergraduate and graduate courses in industrial systems engineering, civil engineering, and sustainability. It has recently paired with private companies in the region to study the feasibility of offshore wind installations on the Georgia coast.

Illinois Institute of Technology
Wanger Institute for Sustainable Energy Research
www.iit.edu/wiser/
3300 South Federal Street
Chicago, IL 60616-3793
(312) 567-3000

Students and researchers at Illinois Institute of Technology (IIT) are striving to improve the quality of life in our nation while preserving the natural resources and the environment for future generations. At the Wanger Institute for Sustainable Energy Research (WISER), more than 50 faculty members and their students are currently involved in energy and sustainability research and educational activities across the colleges and institutes at IIT.

Iowa Lakes Community College
www.iowalakes.edu
300 South 18th Street
Estherville, IA
(712) 362-2604 or (800) 521-5054

Iowa Lakes Community College's Wind Energy and Turbine Technology Program offers a two-year associate in applied science program that helps meet a growing demand for skilled technicians who can install, maintain, and service modern wind turbines. The diploma program consists of three terms of coursework, providing training in the construction, maintenance, and operation of wind turbines. A second year of coursework leads to the associate's degree, with additional training in the diagnosis of turbines, computerized control and monitoring systems, wind turbine siting, and data acquisition.

Iowa State University
www.windenergy.iastate.edu/reu.asp
Wind Energy Science, Engineering, and Policy (WESEP)
College of Engineering
104 Marston Hall
Ames, IA 50011
(515) 294-1588

Iowa State University offers an intensive ten-week on-campus research program in Wind Energy Science, Engineering, and Policy (WESEP) for undergraduate students. Ten fellowships are sponsored each year by the National Science Foundation's (NSF) Research Experiences for Undergraduates (REU) program. Students work collaboratively in interdisciplinary teams with faculty and graduate students to receive training and get hands-on research experience in areas that address critical, long-term national needs in wind energy-related areas.

MIT Laboratory for Energy and the Environment
http://web.mit.edu/mitei/lfee/
Contact: Teresa Hill, Ph.D.
MIT Room E19-370U
Cambridge, MA 02139
(617) 253-1341

The Laboratory for Energy and the Environment (LFEE), an integral part of the MIT Energy Initiative (MITEI),

fosters collaboration among industry, government, academia, nongovernmental organizations, and the public to address not only the complex interrelationships between energy and the environment, but also the technological, economic, and social aspects of sustainable energy development and use. Graduate fellowships at MIT are vital support for MIT's interdisciplinary research in energy, the environment, and sustainability topics.

Oregon State University
www.oregonstate.edu
Corvallis, OR 97331–4501
(541) 737-1000

Oregon State offers more than two hundred undergraduate and one hundred graduate degree programs through its twelve colleges, including the University Honors College, one of only a handful of degree-granting honors programs in the United States. Numerous programs offered through the College of Engineering pertain to wind energy at both the undergraduate and graduate levels, while other departments and colleges provide training in related disciplines.

Riverland Community College
www.riverland.edu
Albert Lea Campus
2200 Riverland Dr.
Albert Lea, MN 56007
(507) 379-3300

Austin Campus
1900 8th Avenue NW
Austin, MN 55912
(507) 433-0600

Owatonna Campus
965 Alexander Drive SW
Owatonna, MN 55060
(507) 455-5880

Riverland's wind turbine technician

diploma program offers students a new career path within the construction electrician and industrial maintenance and mechanics programs. This option gives students a choice of training to be an entry-level wind turbine technician, someone who plays a key role in ensuring quality, safety and service involving the operation and maintenance of wind turbine units, performing mechanical and electrical troubleshooting as well as repair and preventative maintenance.

Texas Tech University
Wind Science and Engineering
Research Center
www.depts.ttu.edu/weweb/
3301 4th St
Lubbock, TX 79415
(806) 742-2011

Texas Tech University's newly formed National Wind Institute (NWI) is based on a strong foundation of more than four decades of research and education on the impact of wind on structures and human life. TTU has created the NWI to better support the interdisciplinary research and educational opportunities in wind science, engineering, and energy.

University of Massachusetts Center for Energy Efficiency and Renewable Energy
www.ceere.org
160 Governors Drive
Amherst, MA 01003
(413) 545-0684

The Center for Energy Efficiency and Renewable Energy (CEERE) provides technological and economic solutions to environmental problems resulting from energy production, industrial, manufacturing, and commercial activities, and land use practices. The research program is built upon four subgroups with unique abilities to service energy and environmental problems. CEERE offers research, training and educational experiences for graduate and undergraduate engineers and scientists.

University of Strathclyde
www.strath.ac.uk/windenergy
UK Wind Energy Research—Doctoral Training Centre
16 Richmond Street, Glasgow G1 1XQ
Scotland, United Kingdom
44 (0)141 552 4400

The overall aim of the Research Centre is to meet the needs of the fast growing wind energy industry by providing high-caliber PhD graduates with the specialist and leadership skills necessary to lead future developments in wind energy systems. The objectives are to ensure that students from different disciplines gain competencies in core aspects of wind energy systems engineering.

Geothermal Energy

Of all the renewable sources of energy that are available to us, the one that may be the most easily accessible lies beneath our feet. Heat from the earth, called geothermal energy, uses steam and warm water that rise naturally from underground reservoirs. Across much of the planet, the subsurface temperature rises substantially—by as much as 150°F (66°C) in some places—for every mile of descent, so that a pipe run to less than a mile can bring up water warm enough to bathe in and wash the dishes. One method now being explored is to send cold water to a depth of about two miles in an especially hot zone and let the earth heat it naturally; when the water returns to the surface, it does so at a temperature of 460 to 600°F (237–315°C), enough to create plenty of electricity from steam. This indirect use of geothermal energy (GEP or geothermal electric power) is very promising: there's plenty of hot underground water avail-

Geysers are a testament to the power of geothermal energy.

able, enough to supply our energy needs for 100,000 years. Getting to it is a challenge, but then, so is coming up with energy solutions of every other kind.

The water need not even be especially warm to be usable. Just a dozen feet below the streets of Chicago, the water flows at a comfortable 55–60°F (13–16°C). Tapping into it by way of a geothermal heat pump (GHP) brings up water that needs to be heated only a few degrees for use in heating in winter, and that's a perfect temperature to send through flooring and ceiling pipes to cool buildings in summer. These medium-temperature water reservoirs can be used in many ways without much modification, providing direct use of geothermal energy. They are abundant in the Midwest, where GHP technology and other geoexchange systems are most widely employed at present.

Overall, geoexchange systems are more energy-efficient than other systems, which translates into lower energy bills and a reduction in the consumption of fossil fuels. Indeed, a 2009 study by New York University determined that geothermal energy is "the most efficient renewable energy alternative and is improving the fastest."

These exchange systems are also clean. Lake County, California, the largest geothermal field under production in the world, is one of the few counties in the state to meet all federal and state air-quality standards. The county's air quality has improved in recent years, in fact, because hydrogen sulfide that would ordinarily be released naturally into the atmosphere by hot springs and vents is now filtered out at the geothermal plants—reducing hydrogen sulfide emissions by a whopping 99.9 percent!

Geothermal, in short, has everything going for it. Even though the technology can be expensive, the returns come immediately, making it a comparatively easy sell for those whose jobs it is to—well, sell it.

Even so, countries such as Japan, Iceland, New Zealand, Italy, Indonesia, and Canada have been at the forefront of geothermal-energy development to date. Though the United States has vast geothermal resources, only one percent of the power used in this country is so produced. As of 2010 the geothermal industry in the U.S. has a capacity of nearly 3,000 megawatts (MW), with power plants operating in five states employing about 5,000 people. According to the Geothermal Energy Association, more than 34,000 jobs would result from a doubling of that capacity, with manufacturing and construction jobs typically corresponding to 6.4 jobs per MW of capacity and 0.74 permanent full-time jobs per MW directly related to power plant operation and maintenance. "If the additional jobs brought on by research, direct use applications, and other geothermal activities are considered, the number of direct jobs would be even greater," adds the GEA, noting that these figures consider employment created by electricity production alone.

These figures are conservative overall. The GEA has estimated that with effective federal and state support, as much as 20 percent of the nation's power needs can be met by geothermal energy sources by 2030. Analysts expect the contribution of geothermal to grow markedly in the coming years: the energy is abundant, easily obtained, and, best of all, inexpensive in the long run compared to most other sources of power. That growth means jobs and many opportunities—though, apart from a few western states, at this writing this vast potential is not being backed by extensive investment in either energy development or the educational and technical infrastructure to sustain it.

Geothermal Quick Facts

Geothermal pumps can be used nearly worldwide. The earth's temperature a few feet below the surface is a relatively constant temperature of about 45–58°F, while air temperatures can fluctuate greatly.

Careers in Geothermal Energy

As with other branches of the renewable-energy industry, geothermal energy requires the services of skilled workers, professionals, salespeople, and auxiliary staff such as secretaries, accountants, attorneys, surveyors, and warehouse workers.

Identifying underground reservoirs is the work of trained geologists, geochemists, hydrologists, and geophysicists, almost all of whom enter the industry after undergraduate and, often, graduate training in the geosciences. Hydraulic engineers, often with training in civil engineering as well as geosciences, are responsible for overseeing the drilling of those reservoirs. Environmental engineers and biologists are charged with preparing environmental-impact studies of areas under consideration for development. Mechanical and electronic engineers, chemists, and materials scientists are also involved in researching and developing new and improved geothermal-energy technologies.

In constant demand are geophysicists, who specialize in exploration techniques that use seismic, electromagnetic, and other sensing methods in the search for oil, gas, minerals, water, and the like. Says one job description: "A geophysicist interprets seismic data, and recommends drilling prospects and processing techniques. A variety of equipment and modeling methods are used to prepare maps (structure, contour, isopach and others) that provide essential information for reservoir development and forecasting. The data is also used to support plant operations and comply with governmental reporting." An entry-level position requires at least a bachelor's degree in physics or geophysics, with graduate training preferred.

The direct-use technologies employed by geothermal energy require workers trained in heating and air-conditioning systems, as well as in the building trades. The interface with geothermal energy and electrical systems requires electrical technicians, electricians, electrical machinists, welders, mechanics, and other skilled workers involved in constructing, operating, and maintaining power plants. Secondary installations, such as geothermally heated pools and spas and radiant flooring, also require trained technicians and installers.

Mechanical engineers, geologists, drilling crews, and heating, ventilation, and air conditioning contractors are needed to manufacture and install GHPs. (Less maintenance is required for geoexchange systems than other technologies, since most of the equipment involved is not exposed to the elements.)

According to the International Ground Source Heat Pump Association (www.igshpa.okstate.edu), an accredited GHP vertical loop installer will be trained in the following subjects:

- GSHP System Design and Layout Basics
- System Materials
- Pressure Drop Calculations
- Thermal Conductivity
- Drilling Processes
- Containment Procedures

- Grouting Concepts
- Air and Debris Purging
- Pipe Joining Techniques
- Project Bidding
- Partnerships

At a specialized level, a geothermal heat pump (HVAC) engineer troubleshoots, repairs, and maintains the whole system: pipes, motors, electrical components, generators, and power transfer switches. Here is a job description posted on one site in the so-called Tech Valley of upstate New York, where geothermal energy is now being developed:

A Geothermal Heat Pump HVAC Engineer must continuously evaluate HVAC operation, performance and documentation requirements, identify specific improvement needs, and provide recommendations or implement system upgrades or improvements. The Engineer may be required to coordinate contractor personnel who perform repairs, modifications, and installation of HVAC and refrigeration equipment. He or she must have the ability to work on any repair or project as needed, independently or as a team member. He or she

A worker monitors equipment at The Geysers geothermal power plant, one of the largest in the world. Photo: David Parsons, NREL

must work with a variety of hand tools, measuring devices, power tools, milling machines, lathes, band saws, welding equipment, bench grinders, and materials. The ability to perform minor electrical repairs, along with ability to read electrical schematics and to detect and report improper operations, faulty equipment and defective machines and unusual conditions to proper supervision is also required. He or she must perform repairs to AC systems and air handlers and create and maintain accurate service logs for a variety of building equipment including water source heat pumps, cooling tower, and other equipment.

GHP engineers, the site goes on to remark, are "in high demand and can expect large career growth possibilities. With a degree and experience, management and supervisory positions can be obtained." Entry-level positions start at around $30,000; a worker with five years of experience can expect to make $40,000–$50,000. For more on the training required for technical positions, consult the "geothermal training" section of HeatSpring Learning Institute's website (*www.heatspring.com*), which offers online courses and certification in a range of fields related to geothermal energy.

At this writing, with a faltering national economy, geothermal energy regimes are not expanding as vigorously as exponents have hoped and projected. In 2011, only some 3,000 jobs were created in the geothermal power

Amedee Geothermal Venture I power plant in Wendel, California. Photo: J.L. Renner, INEEL

and commercial/residential GHP fields, almost all of them in California and Nevada, where geothermal energy is a substantial part of the renewable energy mix. However, the Geothermal Energy Association predicts stronger job growth in the near future as the economy recovers and investors return to the field.

. . . .

How Geothermal Energy Works: Geoexchange

Geoexchange (sometimes called geothermal, or ground-source heating and cooling) taps the renewable, safe, and virtually endless energy supply that lies just below the earth's surface.

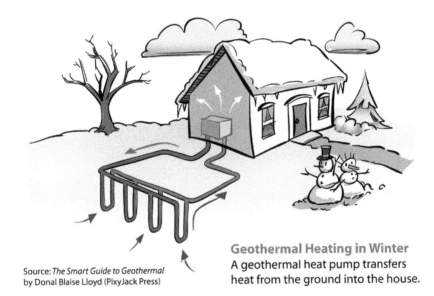

Geothermal Heating in Winter
A geothermal heat pump transfers heat from the ground into the house.

Source: *The Smart Guide to Geothermal* by Donal Blaise Lloyd (PixyJack Press)

In winter, warmth is drawn from the earth through a series of pipes, called a loop, installed beneath the ground. A water solution circulating through this piping loop carries the earth's natural warmth to a heat pump inside the home. The heat pump concentrates the earth's thermal energy and transfers it to your home. In the summer, the process is reversed; heat is extracted from air inside the house and transferred underneath the surface by way of the ground loop piping. The geoexchange system also uses some of the heat extracted from the interior in the summer to provide free hot water—saving as much as 30 percent on your annual hot water bill.

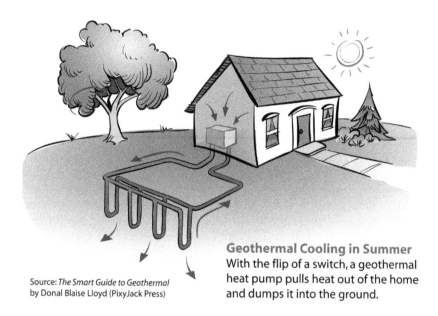

Geothermal Cooling in Summer
With the flip of a switch, a geothermal heat pump pulls heat out of the home and dumps it into the ground.

Source: *The Smart Guide to Geothermal* by Donal Blaise Lloyd (PixyJack Press)

Because geoexchange technology uses such a readily available source of energy—and uses it so efficiently—it can save a substantial amount of money on monthly utility bills. In fact, a typical 1,500-square-foot home in a moderate climate can be heated and cooled for a year-round average of just $1 a day. *Source: Geothermal Heat Pump Consortium*

Drilling a well for a closed-loop geothermal system at NREL's campsu. Photo: Devin Egan

Knowledge Is Power

The preceding position for a GHP engineer requires a high-school diploma or the equivalent and appropriate HVAC certifications. "A two-year degree and bachelor's degrees for advancement/supervisory positions may be preferred but not always required," the advertisement continues.

Other positions within the geothermal-energy field require a degree from a community college or four-year college. One especially promising program, inaugurated in the fall of 2007, is located at the University of Nevada at Reno, where, in concert with the Truckee Meadows Community College, students have the vast Steamboat Geothermal Complex nearby as a field laboratory. UNR also offers a minor in renewable energy through its Department of Civil and Environmental Engineering, describing the course of study as follows:

> The objective of the interdisciplinary minor is to provide students with technical skills, economic and political background, and analysis and design skills that will help them to better apply the knowledge gained in their major to the nationally important issues of alternative and renewable energy.
>
> Students will be exposed to a broad range of technical and social/political disciplines necessary to understand the sources of renewable energy, technical and economic decisions involved in using alternative energy sources, and the policy and regulatory issues that influence adoption of alternative energy resources.

UNR is also the flagship institution for the recently founded National Geothermal Academy, a consortium of schools that include Southern Methodist University, West Virginia University, Cornell University, Stanford University, and the University of Utah, all considered top schools in the geothermal field. The NGA offers an intensive eight-week summer program that yields six units of credit; for more information, see the resources at the end of this chapter.

Other schools with long-established programs in "old" energy—that is, fossil energy, through such courses of study as petroleum engineering and exploration—are developing parallel programs in renewable energy, though these sometimes do not enjoy equivalent status. For example, the Colorado School of Mines, one of the nation's leading mineral-exploration centers of study, now offers a minor in energy, with "three curricular tracks: Fossil Energy, Renewable Energy, and General." CSM's catalog continues:

All Energy Minors must take Introduction to Energy, ENGY200, and Energy Economics, EBGN330/ENGY330, and Global Energy Policy, ENGY490. In addition to the required courses, students in the Fossil Energy track must take ENGY310, Introduction to Fossil Energy, and two approved fossil energy-related electives. In addition to the required courses, students in the Renewable Energy track must take ENGY320, Introduction to Renewable Energy, and two approved renewable energy-related electives.

Workers within the renewable energy field generally can benefit from workshops and conferences (see resources at the end of this chapter), as well as advanced training. One recent development is the Association of Energy Engineers (AEE) Certified GeoExchange Designer (CGD) program, which, AEE says, is "designed to recognize professionals who have demonstrated high levels of experience, competence, proficiency, and ethical fitness in applying the principles and practices of geothermal heat pump design and related disciplines, as well as to raise the professional standards within the field, and to encourage those involved in the design process

Pipelines to a geothermal plant.

through a continuing education program of professional development." Eligibility for the program is contingent on meeting one of the following sets of requirements:

A four-year engineering degree and/or P.E. [Professional Engineer], and/or R.A. [Registered Architect] with at least three years of combined experience in commercial geothermal heat pump design, and/or heating, ventilating and air conditioning field; OR

A four-year non-technical degree, with at least five years of combined experience in commercial geothermal heat pump design, and/or heating, ventilating and air conditioning field; OR

A two-year technical degree, with at least eight years of combined experience in the geothermal heat pump design, heating, ventilating and air conditioning field; OR

Ten years or more of verified combined experience in commercial geothermal heat pump design, and/or heating, ventilating and air conditioning field.

All candidates for certification must take an online exam. Those who pass the test can expect many opportunities for professional advancement—and lots of work in the years to come. But at every level—white collar, blue collar, and green collar—geothermal energy holds plenty of promise for everyone. ❖

· ·

Blue Sky Energy Solutions

David Petroy and business partner Eldred Himsworth design commercial and residential geothermal systems in Colorado (*www.bluesky-energy.com*). David shares his thoughts:

What motivated you to launch Blue Sky Energy Solutions? *I've been involved in energy industry throughout my professional career and have always had a personal passion for the outdoors and nature. Launching a company that helps society adopt new and better ways to provide the power needed to function satisfies my scientific and engineering interests in energy and helps lessen the impact of fossil fuels on the environment.*

For people wanting to enter this field, what education and skills do you recommend? *For system designers I would recommend a mechanical engineering degree or a building-energy engineering degree. For applications engineers, good communication skills are essential to understanding the client's needs and for presenting appropriate options. Both designers and engineers should study energy economics and/or micro economics, and a course or two in hydrogeology and/or near-surface geology would be very helpful. And everyone should have a basic course in business.*

For installers, certification in excavating equipment operation and a commercial truck license is required. We will need many more well drillers if the industry is going to expand. This is a specialized field which requires drilling-training certification and apprentice experience for a few years.

Do you see growth potential in this industry? *The geothermal (ground source heat pump) heating industry has been growing slowing and steadily over the past 20 years through all types of economic conditions. It is a well-proven, solid technology. The work is rewarding and every job has a slightly new twist—which keeps things interesting.*

Slinky coils, prior to burial, for the home's GHP system. Photo by GeoSystems, LLC

Resources

Organizations & Programs

American Ground Water Trust
www.agwt.org
(603) 228-5444
AGWT's geothermal seminars, which are geared to potential end-users and to professionals who design, install, inspect, approve, recommend or regulate geothermal systems, would be an informative place to get up to speed with this technology.

Association of Energy Engineers
www.aeecenter.org
(770) 447-5083
A nonprofit professional society of 8,500 members whose mission is to promote the scientific and educational interests of those engaged in the energy industry and to foster action for sustainable development. The society publishes useful information on its web site, and it offers a range of scholarships in energy and management.

Environmental and Energy Study Institute
www.eesi.org
(202) 628-1400
EESI describes its mission as "educating Congress on energy efficiency and renewable energy; advancing innovative policy solutions." It offers general information to interested individuals on renewable energy matters.

Geothermal Energy Association
www.geo-energy.org
(202) 454-5261
GEA is a trade association composed of U.S. companies that support the expanded use of geothermal energy and are developing geothermal resources worldwide for electrical power generation and direct-heat uses. Its web site offers a wealth of information on all aspects of geothermal energy, as well as information on its annual trade show, which draws exhibitors and participants from all over the world.

Geothermal Heat Pump Consortium
www.geoexchange.org
(202) 558-6759
Another trade association, the GHPC is a good source of information about geoexchange technologies, professional licensing, and other topics.

Geothermal Resources Council
www.geothermal.org
(530) 758-2360
GRC members represent a broad spectrum of geothermal professionals and companies from around the world. They serve as the focal point of continuing professional development for its members through many outreach, information and technology transfer, and educational services. Their web site also provides easy access to the International Geothermal Association.

Geothermal Technologies Program
www1.eere.energy.gov/geothermal
This U.S. Department of Energy site offers news on educational programs, scholarships, and workshops as well as documents of many kinds on geothermal energy.

HeatSpring Learning Institute
www.heatspring.com
(800) 393.2044
An education company located in Cambridge, Massachusetts, that focuses on clean energy training. They offer IGSHPA geothermal installer training as well as a 6-week online entry-level, non-IGSHPA training course.

International Ground Source Heat Pump Association

www.igshpa.okstate.edu
(800) 626-4747
IGSHPA is a non-profit, member-driven organization established to advance geothermal heat pump technology at local, state, national and international levels. They conduct training on the campus of Oklahoma State University. Traveling workshops are also available. They also produce publications such as Ground Source Installation Standards. Successful completion of the training and the exam for the Accredited Installer Workshop or the Accredited Vertical Loop Installer course provides an IGSHPA installer's card, a certificate, a complete set of manuals and membership in the IGSHPA.

Publications

Geothermal Energy: Renewable Energy and the Environment by William E. Glassley (CRC Press, 2010). This extensive—and expensive—textbook reviews the background, theory, power generation, applications, strengths, weaknesses, and practical techniques for implementing geothermal energy projects on a large scale. The book covers geosciences principles, exploration concepts and methods, drilling operations and techniques, equipment needs, and economic and environmental topics.

Geothermal Energy: An Alternative Resource for the 21st Century by Harsh K. Gupta and Sukanta Ray (Elsevier, 2006). This book studies various facets of geothermal energy development and summarizes the current knowledge of geothermal resources and their exploitation. Accounts of geothermal resource models, various exploration techniques, and drilling and production technology are discussed.

Geothermal Heat Pumps: A Guide for Planning and Installing by Karl Ochsner (Earthscan, 2007). Ochsner, a European specialist in geothermal systems, introduces basic theory and reviews the wide variety of available technology for homes and businesses. The book offers information on planning and system control, using data, graphics, and tables from the worldwide market.

Residential Geothermal Systems: A Guide to Using the Ground Below by John Stojanowski (Pangea, 2011). Through this technical publication, readers will learn about the design and functioning of heat pumps and their deployment in extracting heat from relatively low temperature water circulating in ground loops. The publisher adds, "They will also learn how to estimate the size of the heat pump required and the ground loop size as well for straight 2-pipe, 4-pipe, 6-pipe and Slinky loop configurations."

The Smart Guide to Geothermal by Donal B. Lloyd (PixyJack Press, 2011). Although written for homeowners this book contains a good deal of information that can be useful to anyone considering a career in the residential geothermal heat pump market, from understanding the various technologies to working with customers.

Where to Study

Boise State University

http://earth.boisestate.edu
Department of Geosciences
1910 University Drive
Boise, Idaho 83725-1535
(208) 426-1631

The Department of Geosciences of Boise State University, located in Idaho's capital, describes its program as follows: "Our research seeks not only to advance understanding of the surface, near surface, and deep Earth environments, but also to produce science that addresses societally relevant problems such as climate change, human-environment interactions, alternative energy sources, and basic materials." Boise State is also the center of the federally funded National Geothermal Data System (NGDS), offering opportunities for internships and graduate-level research.

Colorado State University

www.warnercnr.colostate.edu
Warner College of Natural Resources
101 Natural Resources Building
Fort Collins, CO 80523–1401
(970) 491-6675

Warner College of Natural Resources is committed to offering a comprehensive range of undergraduate and graduate degree programs that directly address today's most important environmental and natural resource issues. Its programs are grounded in state-of-the-art science and technologies and involve students in direct problem-solving experiences. Its students are well prepared to become leaders in environmental and natural resources management and science.

Cornell University

www.geo.cornell.edu/eas/energy
College of Engineering
Ithaca, NY 14853
(607) 255-5241

Cornell University is a leading center for the study of renewable energy. Its College of Engineering offers minors for undergraduate and graduate students in Sustainable Energy Systems, including geothermal energy.

Montana Tech of the University of Montana

www.mtech.edu
1300 West Park Street
Butte, MT 59701
(800) 445-8324

Offering two- and four-year degrees in both trade and academic fields, Montana Tech has a well-regarded program in geological and geophysical engineering and is located in one of America's most active areas for geothermal energy.

Oklahoma State University

www.hvac.okstate.edu
Stillwater, OK 74078
(405) 744-5000

The Building and Environmental Thermal Systems Research Group of Oklahoma State University is made up of faculty members, students, and researchers with interests that include building heat transfer, HVAC systems modeling, building energy simulation, hydronic heating systems, geothermal heat pump systems, and ground loop heat exchanger technology.

Oregon Institute of Technology

www.oit.edu

With campuses in Klamath Falls, Wilsonville, and La Grande, as well as an out-of-state extension in cooperation with Boeing Aircraft in Seattle,

OIT is home to the Geo-Heat Center, a research institution that "provides technical analysis for those actively involved in geothermal development." OIT also offers an undergraduate renewable energy degree, which includes a course in geothermal energy and ground-source heat pumps. As of 2012, many of its courses were available online.

Rensselaer Polytechnic Institute
www.rpi.edu/dept/ees/
Department of Earth and
Environmental Sciences
Jonsson-Rowland Science Center, 1W19
110 8th Street
Troy, NY 12180
(518) 276-6474

RPI offers undergraduate and graduate training in the geosciences and has become a leading center of geothermal-energy studies. The programs include the study of Earth's component materials, the development of its structures and surface features, the processes by which these change with time, and the origin, discovery, and protection of its resources—water, fuels, and minerals.

University of California, Merced
naturalsciences.ucmerced.edu
5200 North Lake Road
Merced, CA 95343
(209) 228-4400

Merced's undergraduate major in Earth Systems Science is designed to provide students with a quantitative understanding of the physical, chemical, and biological principles that control the processes, reactions, and evolution of Earth. Core courses cover the fundamentals of chemistry, biology, hydrology, ecology, and earth sciences. Graduates are well prepared for either graduate studies or jobs in the areas of environmental science and conservation, ecosystem and natural resource management and science, and many aspects of agricultural sciences.

University of Nevada, Reno
www.unr.edu
1664 N. Virginia Street
Reno, NV 89557–0042
(775) 784-1110

A land-grant institution with graduate programs in a wide range of fields, Nevada has a well-established program of study in renewable energy. The program includes students from Truckee Meadows Community College, using the vast, nearby Steamboat Geothermal Complex and Nevada's abundant sunshine as laboratories. For information on the specialized program of study under the rubric of the National Geothermal Academy, see www.unr.edu/geothermal/NGA.htm.

University of Wyoming
www.uwyo.edu
School of Environment and Natural Resources
P.O. Box 3963
Laramie, WY 82701-3963
(307) 766-5080

The University of Wyoming School of Environment and Natural Resources provides degree programs in research and policy aspects of environmental and natural resource decision-making. University of Wyoming focuses on six operational clusters: energy; ecosystems; allocation systems; environmental conditions and economic and social interactions; land and water resources; and atmospheric resources and systems. The University's School of Energy Resources also provides technical education in both renewable and nonrenewable energy regimes.

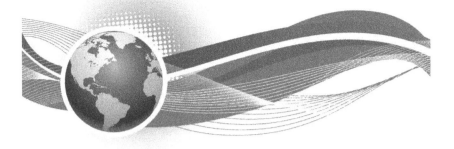

Hydropower

Hydroelectric Power

You hear it long before you see it, a dull roar shrouded in steam, hidden from view. Then you see it: Horseshoe Falls, where the Niagara River drops 180 feet (55 meters) into Niagara Gorge, on the border between the United States and Canada. Between May and November, when the river is usually free of ice, 675,000 gallons of water tumble over the falls each and every second.

On the other side of the continent, the states and provinces of the northern Pacific Coast have long relied on water-generated electricity.

Today there are some 160 hydroelectric facilities in the region, providing Washington State with nearly 80 percent of its power and Idaho nearly all of its supply. The biggest of these facilities is the Grand Coulee Dam. California and Oregon also have numerous facilities.

Hydroelectricity is the most prevalent form of renewable energy throughout the nation, accounting for about 90 percent of the total of renewable resources. In 2009, Energy Secretary Steven Chu praised hydropower as the nation's "lowest-cost energy option," and the statistics bear him out. Meanwhile, the National Hydropower Association urges that those statistics "demonstrate the industry's extraordinary potential to expand its contribution to the country's energy, environmental, and economic goals: Hydropower can create 1.4 million cumulative jobs and add 60,000 megawatts of affordable, domestic, renewable energy by 2025."

Most of the nation's dams—and there are about 80,000 of them—do not generate electricity, but many are now being studied for the possibility of retrofitting them to produce power. Added to the mix of power, hydroelectricity is clean and abundant. It is also inexpensive relative to other forms of power, and as long as dams and plants are engineered in such a way that the health of rivers and their wildlife populations is assured, there are few reasons not to put water to greater use in providing power, at least in places where water is abundant. In that spirit, current engineering research is focusing on building turbines that do not require damming and that allow fish to pass through them without harm.

Ice Harbor Dam near Burbank, Washington. Photo: US Army Corp of Engineers

How Hydropower Works

Mechanical energy is derived by directing, harnessing, or channeling moving water. The amount of available energy in moving water is determined by its flow or fall. Swiftly flowing water in a big river, like the Columbia River along the border between Oregon and Washington, carries a great deal of energy in its flow. So, too, does water descending rapidly from a very high point, like Niagara Falls in New York.

In either instance, the water flows through a pipe, or penstock, then pushes against and turns blades in a turbine to spin a generator to produce electricity. In a run-of-the-river system, the force of the current applies the needed pressure, while in a storage system, water is accumulated in reservoirs created by dams, then released when the demand for electricity is high.

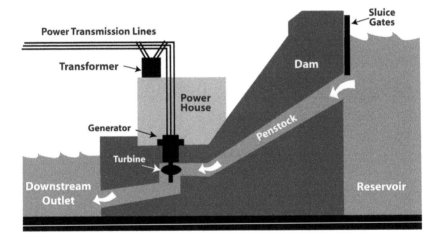

Meanwhile, the reservoirs or lakes are used for boating and fishing, and often the rivers beyond the dams provide opportunities for whitewater rafting and kayaking. Hoover Dam, a hydroelectric facility completed in 1936 on the Colorado River between Arizona and Nevada, created Lake Mead, a 110-mile-long national recreational area that offers water sports and fishing in a desert setting. *Source: U.S. Department of Energy*

Tidal (Marine) Power

There is also exciting research being undertaken on a still larger, inexhaustible source of power—namely, oceanic tides, which can turn massive turbines and provide abundant energy to vast portions of the world. Only a couple of dozen places on Earth have natural inlets and a large enough tidal range—about 10 feet (3 meters)—to produce energy economically, but that has more to do with our technology than with the potentials of nature.

There are currently only a few commercial tidal-energy dams in operation, in places such as France, England, and Canada. Given that more than 60 percent of the world's human population lives within 30 miles of a coastline, bringing wave power to bear as an energy source seems an eminently logical thing to do, but scientists since the time of Leonardo da Vinci have been thwarted in their efforts to produce robust enough generators and turbines that the sea does not soon destroy. Usually, those generators and turbines have been located against shorelines, capturing waves as they crash onto land. The toll on machinery is consequently heavy, and the output from that machinery has not been efficient, particularly because there is no surefire way made to connect tidal energy plants to the larger electrical grid.

That is changing, however, as the technology is maturing. A recently constructed tidal plant at Strangford Lough, on the coast of Northern Ireland, operates like an underwater windmill, with free-turning rotor blades as opposed to the comparatively inefficient fixed-direction ones of earlier designs. Under usual tidal conditions, the plant's two rotor blades can each power about 1,000 homes.

That's not much, but the Strangford plant serves as a template for smaller tidal-power plants, and its design is being adapted to other coast-lines on the North Sea. Meanwhile, the South Korean coast is the site of another prototype whose design resembles a motorboat's propellers rather than a windmill. At a larger scale, it is projected to provide enough power for 400,000 households.

But must a turbine be submerged in order to capture wave power? A Scottish scientist named Richard Yemm thinks not. Yemm, who holds a doctorate in mechanical engineering from Edinburgh University, has designed a "sea snake," a bright red, tube-shaped metal device that looks something like a seaborne bullet train floating atop the waves, its encased turbines turning like waterwheels on a stream. An individual snake array cannot yet generate industrial-scale amounts of energy, but one advantage of the design is that it can be inexpensively introduced everywhere that the tide flows, so that municipalities and even oceanfront households might one day be able to harvest their own power from the sea. Given that sea levels are expected to rise worldwide by 1.1 meters by the year 2100, it's a timely innovation.

Indeed, climate change is leading to stormier waters and taller waves; between 1960 and 1990, average wave heights rose by 25 percent, and insurance companies have documented a steady rise in the number of wave-caused calamities. Tidal and wave energy therefore holds extraordinary promise to put the bounty of the earth to good use without causing harm.

Ocean Power Technologies' PowerBuoy® is designed to convert ocean wave energy into useable electrical power for utility-scale grid connected applications. Photo: Ocean Power Technologies

Careers in Hydropower

As with all forms of renewable energy, water power, from the oceans and from the world's rivers alike, needs the minds and efforts of students, researchers, thinkers, and workers from all places and many walks of life. And, as with other forms of renewable energy, hydroelectric power requires a blend of skilled workers, technicians, researchers, operators, engineers, electricians, salespersons, and auxiliary staff. Projected hydroelectric plants must also be assessed for their potential effects on fish and wildlife populations and on river flows, affording opportunities for wildlife specialists, environmental scientists, biologists, hydrologists, and other scientists. Arguments for the construction of new projects—pro and con—require advocates, writers, and legal specialists. Economists and energy analysts study the economic need for new facilities and their effects on business and society.

In short, hydroelectric power welcomes individuals with many skills and interests, with backgrounds in many disciplines and branches of science and engineering. Richard Yemm's firm, Pelamis Wave Power (*www. pelamiswave.com*), puts it thus:

Pelamis Wave Power is committed to working with talented and innovative professionals who are passionate about the work that we do. We employ and support motivated, talented, and dedicated individuals who actively contribute towards the company's growth and development, and are enthusiastic about renewable energy and innovative technology. Pelamis encourages co-operation and team work, and we don't see job titles as labels, with people from multiple engineering disciplines and backgrounds working together to understand how to make the best wave power technology possible.

At the heart of hydroelectric energy development is hydrology, the scientific study of water and its properties. Hydrologists are called on to analyze where water supplies can be found and how long those supplies are likely to last, and no discussion of resource policy is complete without their participation. Early in their careers, hydrologists are likely to spend most of their time in the field, mapping water supplies and pathways, testing water quality, and planning projects on the ground. As they advance, many hydrologists engage in laboratory or office work, often acting as liaisons to government agencies and businesses and analyzing computer-generated projections of water availability and behavior.

According to the U.S. Bureau of Labor Statistics, in 2011 there were about 7,000 positions for hydrologists at all levels in the United States, with

a median annual income of $79,000. It has been projected that the demand for hydrologists will grow slightly throughout the rest of the current decade, mostly with an eye toward monitoring compliance with environmental regulations, as well as working in flood control. Increased emphasis on hydroelectric power would increase opportunities for hydrologists.

Hydraulic engineers and hydraulic technicians work in conjunction with hydrologists to develop and operate hydroelectric plants and related facilities. Civil engineers (average salary $82,700) and structural engineers (average salary $92,000) are involved in building dams and other structures, and electrical engineers and mechanical engineers develop and maintain power systems that are in turn staffed by power-plant operators and other technicians. Most such systems are operated and monitored by computer, requiring skilled information-technology workers (average salary $51,000).

Additionally, many dams and hydroelectric facilities are located within federal parks, national forests, and other parts of the public domain, served by park rangers, forest-service officers, and other officials. In many such venues, there are also opportunities for recreation workers, wilderness guides, and other outdoor-activities specialists.

According to energy analyst Rakesh Radhakrishnan, growth in hydropower jobs will take place in these regions of the United States: (1) Western states with the greatest resource potential (Washington, California, Oregon, and Alaska); (2) manufacturing states (among them Pennsylvania, Wisconsin, Tennessee, and Ohio); and (3) states with advanced hydro potential (including Florida, Maine, New York, and Tennessee). For obvious reasons, states in the Southwest and Great Plains regions are not well represented on this list, but even they will be seeing some growth in hydropower needs and the jobs to serve them.

Hydropower Quick Facts

The U.S. is the second-largest producer of hydropower in the world, behind Canada. Norway produces more than 99 percent of its electricity with hydropower. New Zealand uses hydropower for 75 percent of its electricity.

Knowledge Is Power

Most technical positions in hydroelectric power require an associate's degree at minimum. Here, for example, are the posted requirements for a senior hydroelectric power technician whose principal duties are to repair and maintain the hydraulic, electrical, and mechanical systems of a power plant in the Northeast:

- Must have a minimum of 7 years experience working with complex mechanical, hydraulic, and electrical systems—ideally including turbines.
- Must be able to read and understand electrical, hydraulic, and mechanical drawings and schematics.
- Must have the ability to work well on a team basis.
- Strong verbal and written communication skills are essential.
- Solid problem-solving and organizational ability is a must.
- Must have familiarity with OSHA safety standards.
- Knowledge of basic Windows software including Outlook, Excel, Word, and Access is required.
- Must have a minimum of a two-year degree in Electrical Engineering Technology or a related technical discipline.

The successful candidate for such a job might find himself or herself working alongside or even for a project engineer, whose skill set and education will be substantially different. Here is a posting for that position in the same geographic area:

Work will involve leading and performing the preparation of conceptual, feasibility and preliminary and final design work products. The work requires coordination with clients, self direction and ability to work as an individual or in a group. Activities will include a wide range of water resources development study tasks, such as hydrological studies to estimate river flows and floods, formulation of dam and hydropower project development alternatives, preparation of preliminary layouts of the major project components, estimating quantities of construction materials, assisting in preparation of project cost estimates, assisting in operational studies to evaluate river flow and hydropower production, and economic studies to evaluate alternative projects and to optimize overall project configuration. Activities will be focused on civil engineering project development aspects, in coordination with mechanical and electrical specialists. The work environment is results oriented and focused on preparing reports or other deliverables within established budgets and schedules. Hands-on knowledge and familiarity with modern computation hydrologic and hydraulic computer methods and CAD systems are required.

That job, the advertisement continues, requires a master's of science in civil engineering with an emphasis on hydropower planning, hydrology, hydraulics, and sedimentation, as well as professional engineering registration and, as with the first position, a minimum of seven years' experience.

A beginning hydrologist, in most instances, must hold a bachelor's degree from a scientific program. Typically, a hydrologist will have training in geophysics, chemistry, engineering science, soil science, mathematics, computer science, aquatic biology, atmospheric science, geology, and oceanography—not all of these fields, of course, but a number of them in combination. Economics, public finance, public policy, environmental law, and business law provide helpful background for hydrologists at more senior levels, and it helps to be comfortable writing reports and giving talks or testimony about water issues.

Here, by way of an example, is the undergraduate course sequence for a BS degree in hydrologic science at the University of California, Santa Barbara:

1) Preparation for the Major (1st and 2nd years) Requirement:
1 course Physical Geology
1 course in Physical Geography
4 courses of Science Calculus
1 course of Intro Statistics
1 year of Intro Sci. Chemistry w/lab
1 year of Intro Sci. Biology
1 year of Intro Physics
1 course in History of Public Policy
1 Intro to Env. Studies or American Political Govt.
1 course in Micro, Macro, or Intro Economics

2) Upper-division Units (3rd and 4th years):
All Hydrologic Sciences majors must complete a total of 56 upper-division units broken down in the following:

Area A. Required Courses (25 total units): Geography 112, Geography 116/Earth Science 173, Earth Science or Environmental Studies 168 or 169, Geography or Environmental Studies 144. Environmental Studies 176A, and Environmental Studies 100 or EEMB 120.

Area B. Complete one of the following three emphases (31 units): Biology and Ecology; Physical and Chemical; Policy.

On that note, here is a job listing for a position as a hydropower engineer working for the Federal Energy Regulatory Commission:

As an entry-level staff engineer you would work with an interdisciplinary team of environmental scientists and engineers, to review, analyze, and resolve engineering and environmental issues associated with proposals to construct and operate non-federal hydroelectric projects; including major dams, reservoirs and powerplants.

You would identify and evaluate alternatives to the project proposed and prepare written analyses describing the impacts of the projects and alternatives with emphasis on engineering-related issues including the need for project power, cost/benefit analysis, hydraulic and hydrologic modeling and the interrelationship of project facilities and flows with other environmental resources such as aquatics and recreation. You would propose recommendations to protect and enhance the environment.

You would visit the project locations before authorization to familiarize yourself with the project site, and if the project is existing, to become familiar with its current and proposed operation.

Your position would involve participation in public scoping meetings, and technical sessions with project proponents, state and federal resource agencies, and other affected parties. Some travel would be required. Training opportunities are provided and encouraged.

Qualifications: A bachelor's degree in engineering; Strong verbal, analytical and writing skills; U.S. Citizenship.

Many hydroelectric facilities within the United States are operated by the U.S. Army Corps of Engineers, which offers civilian employees a four-year training program. One instructor in that program suggests that students interested in such a career path study math and physics—but that they also get a good general education to help them with critical-thinking and problem-solving skills.

One need not study in the United States, of course. One intriguing course of study is the international master's degree in hydropower offered by the Norwegian University of Science and Technology (NTNU) in the far northern city of Trondheim. The NTNU catalogue describes the program so:

The degree program, MSc in Hydropower Development, known as the HPD program, is a two-year long international master's program in hydropower planning.

The first year consists of a series of 6 foundational courses and a larger group project where the students apply knowledge from the other courses by conducting a pre-study of the development alternatives in a Norwegian water system. This involves learning how to combine techniques, environment, and economy to secure success.

The final year consists of four compulsory advanced courses in the autumn, while the entire spring semester is dedicated to the master's thesis. Our lecturers have international experience and are recruited from NTNU and the international hydropower environment.

Regarding the master's thesis, there are numerous options, and most students choose their topic within areas where NTNU offers special qualifications such as: planning of a concrete project, hydrology, hydraulics, sediment transport, engineering geology, and plant engineering. It is also possible to write the thesis in connection with projects abroad.

Hydropower Quick Facts

Hydropower is the most efficient way to generate electricity. Modern hydro turbines can convert as much as 90 percent of the available energy into electricity. The best fossil fuel plants are only about 50 percent efficient.

The catalogue describes an interdisciplinary course of study that involves various branches of engineering with geology and hydrology, along with classes in economics and business. Graduates of the program have found jobs with consulting engineering companies, contracting firms, water and energy authorities, power and utility companies, and governmental and nongovernmental institutions working in the areas of international development and cooperation.

To get a sense of the broad variety of studies that shape a career in hydrology and hydropower, as well as to identify strengths and weaknesses, you might want to take the free courses offered by the University Corporation for Atmospheric Research (UCAR). The course home page can be found at *www.meted.ucar.edu/training_course.php?id=7*. Another excellent source of preparation and background is offered by the US Geological Survey at its website "Education—Learn About Water" (*http://water.usgs.gov/education.html*). ❖

The OE Buoy is designed around teh oscillating water column principle. Its turbine captures the energy of the wave and the generator converts this energy to electrical power. Photo: Ocean Energy Limited

Electricity from the Ebb and Flow of Tides

Top: Installation of fish monitoring equipment. Bottom: RITE Project Phase 2 – A Verdant Power Free Flow turbine is being installed into the East River, New York City.

Verdant Power's Roosevelt Island Tidal Energy (RITE) Project is being operated in New York City's East River. The RITE Project incorporates a Kinetic Hydropower System comprised of Verdant Power's 5 meter, 35 kW tidal Free Flow™ turbines, which generate electricity from the ebb and flow tides of the East River. The Project is progressing from an initial demonstration array of six turbines to a complete arrangement of 100-300 turbines. At full capacity the project could generate up to 10 MW, enough to power nearly 8,000 New York homes. *www.VerdantPower.com*

Resources

Organizations & Programs

Hydroworld
www.hydroworld.com
Hydroworld.com is an omnibus site for several Web-based and print publications having to do with all aspects of hydropower.

National Hydropower Association
www.hydro.org
Since 1983, NHA has been dedicated exclusively to advancing the interests of hydropower energy in North America. Its web site has a wealth of information, and well-tended links to other resources. The associated Hydro Research Foundation web site (*www. hydrofoundation.org*) has good information for students, too.

Ocean Energy, California Energy Commission
www.energy.ca.gov/oceanenergy/
An excellent source for links to companies and organizations involved in ocean energy development.

U.S. Bureau of Reclamation
www.usbr.gov/lc/hooverdam/
This web site is devoted to Hoover Dam, one of the nation's best-known hydroelectric facilities. It offers information on dam operation, as well as resources for students and teachers.

US Department of Energy
www.eere.energy.gov/topics/water.html
Working with industry, the Wind and Hydropower Technologies Program pursues research and development in environmentally friendly technologies to maintain the nation's existing hydropower capacity. The US Department of Energy also funds many other research programs in hydropower.

Publications

Hydroelectric Handbook by William P. Creager and Joel D. Justin (Wiley, 1952). Though more than half a century old, this is a widely used reference in the field, covering the theory and practice of hydroelectric power generation.

Microhydro: Clean Power from Water by Scott Davis (New Society Publishers, 2004). Microhydro is the first complete book on the topic in a decade. It covers principles, design and site considerations, equipment options, as well as legal, environmental, and economic factors.

Hydropower Economics by Finn R. Forsund (Springer, 2010). This book examines sustainable alternate energy sources, beginning with modeling hydropower and extending the model to include thermal power and wind power. The book uses various econometric measures, equilibrium metrics, and productivity analyses to analyze and model the optimal use of these alternate energy sources. It derives results on the allocation of the amounts of alternate sources of energy (water, thermal, and wind) required to produce electricity at acceptable levels over time.

21st Century Pocket Guide to Hydropower, Microhydropower and Small Systems, Incentives and Funding, Dams, Turbine Systems, Environmental Impact and Fish Passage, History, Research Projects (US Department of Interior, 2010). This ebook provides comprehensive coverage of all aspects of hydropower, microhydropower, dams, and turbines, with information on everything from the basics, federal incentives and funding opportunities, environmental impact, fish passage, design concepts, practical small hydropower systems, federal research, and more.

Introduction to Hydro Energy Systems: Basics, Technology and Operation by Herman-Josef Wagner (Springer, 2011). This book is a basic and very useful—though very expensive—introduction to the entire realm of hydropower. The target reader is the already trained civil engineer moving into projects involving hydro energy.

Where to Study

Colorado State University
www.warnercnr.colostate.edu
Warner College of Natural Resources
103 Natural Resources Building
Fort Collins, CO 80523
(970) 491-5629

Warner College of Natural Resources is committed to offering a comprehensive range of undergraduate and graduate degree programs that directly address today's most important environmental and natural resource issues. Its programs are grounded in state-of-the-art science and technologies and involve students in direct problem-solving experiences. Its students are well prepared to become leaders in environmental and natural resources management and science.

Columbia University
www.me.columbia.edu
2960 Broadway
New York, NY 10027-6902
(212) 854-1754

A leading university with undergraduate and graduate instruction in every field that touches on renewable energy, Columbia is home to a highly ranked Department of Mechanical Engineering that offers courses in energy sources, materials science, electrochemistry, and energy systems. See also *www.cheme. columbia.edu*

Norwegian University of Science and Technology (NTNU)
www.ntnu.edu/studies/msb1
Department of Hydraulic and Environmental Engineering
attn. Ånund Killingtveit
NO-7491 Trondheim, Norway
(47) 73 59 47 51

Please see the chapter for a description

of the NTNU international master's degree program in hydropower.

St. Cloud State University

www.stcloudstate.edu
Atmospheric and Hydrologic Sciences
720 4th Avenue South
Saint Cloud, MN 56301
(320)308-3144

The hydrology major, leading to a BS degree, focuses on the quantitative study of surface and ground water and provides the background for a variety of entry level job opportunities in industry and government and for those intending to pursue graduate work in the field. Students are strongly urged to consider a minor in a related field such as meteorology, geography/geographic information science, or environmental studies. Coursework in the major involves environmental and earth sciences, chemistry, geology, mathematics, and physics.

University of California—Davis

http://hydscigrad.ucdavis.edu
Hydrologic Sciences Graduate Group
Veihmeyer Hall
Davis, CA 95616
(530) 752-6810

UC Davis houses one of the nation's most esteemed graduate programs in hydrology. The Hydrologic Sciences Graduate Group (HSGG) offers MS and PhD degrees emphasizing physical, chemical, and biological processes that affect the circulation of water and solutes on Earth. The Group offers options in hydrology, hydrogeochemistry, and hydrobiology; within the hydrology option, students can specialize in surface hydrology, subsurface hydrology, water resources management, or irrigation and drainage. Students with training in hydrologic sciences, geology, geophysics, engineering, soil science, biology, chemistry, computer science, fluid mechanics, mathematics, and physics are strongly encouraged to apply for admission to the graduate program. Coursework includes studies in geographic information systems, geology, mechanics, irrigation, fluid dynamics, limnology, environmental law, and many other subjects.

University of California—Santa Barbara

www.es.ucsb.edu/academics/hydo-sci-major/bs
Environmental Studies Program
4312 Bren Hall
UC Santa Barbara CA 93106–4160
(805) 893-2968

Please see the text for a description of the UCSB undergraduate degree in hydrologic sciences.

Vermilion Community College

www.vcc.edu
1900 East Camp Street
Ely, MN 55731
(800) 657-3608

VCC offers a two-year associate's degree in watershed science. Watershed science graduates work for either government agencies (like the USGS and local Soil and Water Districts) or for private companies that collect water information for profit. As hydrologic technicians, they measure stream flows, collect and analyze samples, monitor groundwater, and study the effects of human and natural activities.

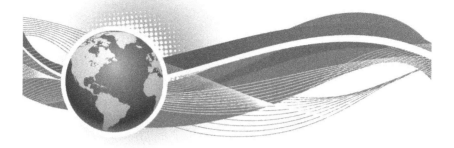

Bioenergy

Can chicken waste help power the world? At a single chicken farm outside Beijing, China, three million hens produce eggs each day—and, at the same time, mountains of waste, 200 tons of it each day. Researchers have put that waste to use, extracting methane that in turn generates electricity.

In the desert of Arizona, fields of blue-green algae are being grown by the ton in large artificial ponds, while trees of the genus *Pongamia* are being grown for oilseed that some analysts believe can be sold profitably for as little as a dollar a gallon.

To the south, in Mexico, agave pulp from which the alcohol called tequila has been extracted serves as a new source of fuel, one more efficient than the fuel produced from corn, sorghum, and other food plants.

Switchgrass grown for biofuels.

Organic matter, or biomass, stores energy. People all over the world have known this fact for a very long time. A tree, for instance, can be thought of as sunlight encased in wood, and the energy it contains—called bioenergy—has been a source of warmth and light for as long as there have been humans. Wood continues to heat our homes, and about one percent of the nation's energy supply comes from the burning of wood waste and other biomass. But things are on the move, and bioenergy shows much promise of becoming one of the leading kinds of renewable energy on the market.

In recent years, owing to state and federal clean-air regulations and other incentives, the petroleum industry has increasingly added bioenergy to nonrenewable fossil-fuel energy, in the form of ethanol. This renewable source, derived mostly from corn in the United States, is far from perfect: at present, under current means of production, it takes lots of fossil fuel to produce ethanol, and, according the Renewable and Appropriate Energy Laboratory at the University of California, ethanol provides only an 18 percent reduction in greenhouse-gas emissions over gasoline. Federal subsidies for ethanol have come under fire in an increasingly austere budget regime, and demand has dropped significantly, just as demand has dropped for the gasoline to which ethanol is added. As the nonprofit group Energize Iowa *(http://energizeiowa.org)* observes, "The subsidy topic is not only a practical one that concerns keeping biofuel production moving forward; it is also a highly political one that will be keeping the agricultural community, consumers, the petrochemical fuel industry, and politicians on edge in the near future."

Other sources are more efficient than corn ethanol. Brazil, for instance, has been producing fuel from less costly, more sustainable sugarcane since the 1970s and has attained its goal of achieving energy independence in the bargain; Brazilian cars run on what is locally called *álcool*, which has the added virtue of having an octane rating of 113, practically putting every driver behind the wheel of a race car, as well as reducing greenhouse gas emissions by up to 70 percent. Trees and grasses, which are called cellulosic biomass, can also be used to make bioethanol, and some fast-growing varieties such as switchgrass, bamboo, hemp, and kenaf are particularly useful—and reduce greenhouse gas emissions to almost zero. Other sources include agricultural residues such as the stalks, leaves, and husks of corn plants; forestry wastes such as chips and sawdust from lumber mills, dead trees, and tree branches; municipal solid waste, including household garbage and paper products; and food processing, papermaking, and other industrial wastes.

Another biofuel—biodiesel—is a mixture of fatty-acid alkyl esters made from vegetable oils, animal fats, and recycled greases. Biodiesel can be used directly in modified diesel engines (and in some newer, unmodified diesel engines), but it is usually used as a petroleum diesel additive. Biodiesel burns fairly cleanly, even if people sometimes joke that vehicles that use it smell like french fries. It burns so cleanly, in fact, that it promises to reduce airborne particulates, carbon monoxide, and hydrocarbons significantly in the coming years. In the United States, most biodiesel is made from soybean oil or recycled cooking oils. In Tucson, Arizona, for instance, several Mexican restaurants contribute their used vegetable oil to a biodiesel program and even allow individual consumers to bring in used cooking oil to add to the mix.

Still another source of bioenergy is methane recovery, as those Chinese researchers have proven. Microorganisms produce biomass gas as they decompose, as do organic materials such as garbage, wood chips, grass clippings, and seaweed. This biomass gas contains methane and carbon dioxide, and it can be collected at landfills and purified to isolate the methane, which can then be used as fuel. Scientists have long been interested in the industrial possibilities of extracting methane from another abundant source—namely, manure, which produces huge amounts of it—and indi-

Current ethanol production is primarily from the starch in field corn kernels, but NREL researchers in the DOE Biofuels Program are developing technology to produce ethanol from the fibrous material in the corn stalks and husks, or other agricultural or forestry residues. Photos: Warren Gretz, NREL

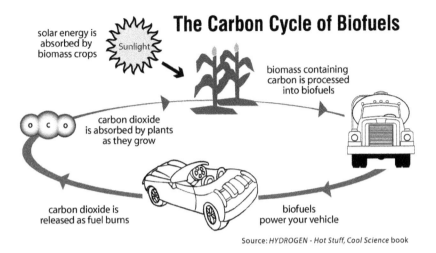

The Carbon Cycle of Biofuels

solar energy is absorbed by biomass crops

Sunlight

biomass containing carbon is processed into biofuels

carbon dioxide is absorbed by plants as they grow

carbon dioxide is released as fuel burns

biofuels power your vehicle

Source: *HYDROGEN - Hot Stuff, Cool Science* book

vidual energy farmers, so to speak, have been doing so at a smaller scale for several years now, even if the whole biogas enterprise has a Mad Max feel to it.

Agricultural researchers, chemists, and other scientists are actively seeking still more sources of fuel, as well as simpler methods of converting biomass into fuel—a process that is now expensive and complicated. Newer fuel blends such as biobutanol are more efficient than bioethanol, and some forms of ethanol produced in this country, such as those from the *Miscanthus* genus of grasses, are better than others, but they still need a little more oomph in order to be truly competitive—and to match that Brazilian potion.

Even so, bioenergy is second only to hydropower as the largest source of renewable energy in the world. Biomass is increasingly taking the place of petroleum in the manufacture of goods such as chemicals, paints, and construction materials. Despite dips in the market, the product of the long worldwide recession that set in in 2008 and that has been agonizingly slow to shake off, bioenergy would seem to have a bright future. This is especially true in the United States, where, according to projections gathered by the Environmental Defense Fund (*www.edf.org/energy/growth-bioenergy*), "Biofuels are projected to account for 60% of the projected growth in total U.S. liquid fuel consumption by 2035." Just so, a new plant on the big island of Hawaii, operated by Hu Honua Bioenergy LLC (*www.huhonua.com*), shows how many jobs are created both directly and indirectly by the growth of bioenergy:

Construction will require approximately 100 workers for at least one year. Fully staffed, the plant will have 28–30 employees. It is projected that more than 100 jobs will be created indirectly in growing and harvesting biomass, in transporting it to the plant, and in other support services. Hu Honua already uses local businesses for services, repairs and supplies and will continue to do so.

Career Avenues

Bioenergy requires, at the outset, men and women who are skilled at making biomass: farmers, gardeners, foresters, growers of all kinds. It requires people who aren't afraid of getting their hands dirty—or greasy.

Apart from that, the bioenergy field requires a broad range of blue-, white-, and green-collar workers, from sales and support staff to truck drivers and warehouse workers to microbiologists, chemists, and biochemists. Engineers, plumbers, and construction workers are needed to design and build bioenergy plants, while electrical and mechanical technicians, mechanical, electrical, and chemical engineers, mechanics, and heavy-equipment operators are needed to run and maintain them. And researchers are needed to identify and develop new, more efficient sources of bioenergy.

To give you a better idea of that range, the U.S. Department of Agriculture hosts a web site devoted to jobs at the Agricultural Research Service (see the resources section at the end of this book). The range of positions listed there shows just how many careers the bioenergy revolution is capable of launching. So, too, does this announcement from a single firm that does bioenergy work in the United States and Europe, listing these job openings in a single week:

Mechanical Piping Engineer
Architectural/Civil/Structural Engineer
Aspen Specialist
Business Development Project Manager
Civil/Structural Engineer
Commodity Manager
Cost Estimation Specialist
Electrical Engineer
Engineering & Construction Manager

Ethanol Sales Coordinator
Hydrolysis & Fermentation Engineer
Instrument/Control Engineer
Mechanical Engineer
Operability & Startup Manager
Pretreatment & Fractionation Engineer
Process Engineer
Process Engineering Project Manager
Project Director
Senior Bulk Material Handling Specialist
Senior Process Engineer
Senior Separations Specialist

Most of those positions are highly specialized, requiring a broad range of skills and training. One of them, that of bioenergy process engineer, for instance, "develops environmentally friendly processes for fractionating plant biomass from crop residues, such as wheat straw, rice straw, rice hulls, and other under-utilized agricultural fibers. The ultimate aim is to produce predictable compositions for use in the creation of composites,

Biodiesel can be made from any fat or oil. Current U.S. biodiesel production is primarily from oil from soybeans such as these, or from recycled restaurant cooking oil. Cleaner burning and renewable biodiesel is most often blended at 20 percent with petroleum diesel. Photos: Warren Gretz, NREL

nanocomposites, and value-added products, such as specialty chemicals and/or ethanol."

Another position at another firm, that of biomass industry development coordinator, is more green-collar-oriented, involving a blend of technical and sales and communications skills. The coordinator, the firm specifies, "gathers information on competitors, prices, sales, and methods of marketing and distribution to enhance success; using survey results to create a marketing campaign based on regional preferences and buying habits. Business (economic resource) and implementation planning is undertaken by either the individual coordinator or their represented organization to assure viability. The coordinator will solicit business from a variety of prospective commercial customers of varied size and industry and will assist in the assessment of data and methods of service delivery."

Finally, many positions are available at the research level, most requiring specialized training and an advanced degree. Here, for instance, is a listing for entry-level fellowships in bioenergy at the Great Lakes Bioenergy Research Center, which suggests a heady set of skills:

The Great Lakes Bioenergy Research Center (GLBRC) at Michigan State University has an opening for an ecologist to investigate the consequences of cellulosic bioenergy production systems for biodiversity and ecosystem services.

The successful candidate will design and direct studies examining the impacts of cellulosic biomass crops on insect biodiversity and ecosystem services at the landscape scale. In addition, the position will hold responsibility for helping to coordinate a team of researchers examining similar impacts on plant and microbial diversity.

We are particularly interested in applicants with excellent organizational and communication skills and some combination of experience in landscape ecology, spatial analysis, GIS, and evaluation of insect-mediated ecosystem services including biological pest control or pollination.

Similarly, a recent listing for two postdoctoral positions at Texas A&M University, a leading center for biofuels research, calls for rigorous skills:

One to two postdoctoral researcher positions in biochemistry and biofuel research will be available at the Institute for Plant Genomics and Biotechnology and Department of Plant Pathology and Microbiology in Texas A&M University. The positions will be part of our expanding plant metabolic engineering group and bioenergy research group. For the first position, the new postdoc researchers will participate in enzyme improvement using HDX mass spectrometry and metabolic engineering. The research will heavily involve protein expression, enzyme assays, pathway optimization, protein modification, and biochemical analysis. The second position may be available for microbial engineering and biomass conversion platform development. We will expect the first postdoc to have solid background in biochemistry, in particular, for protein expression and enzyme assay. Biophysics background will be a plus but not needed. For the second position, we expect the postdoc to have strong background in microbial engineering and fermentation. Engineering background will be a plus but not required.

NREL researchers work on the thermochemical pretreatment step of converting lignocellulosic biomass to fuel ethanol and other valuable chemicals. Photo: Warren Gretz, NREL

Knowledge Is Power

The jobs listed above require a doctorate, as do many research posts, particularly within the university. The bioenergy process engineer position calls for a recent master's degree in engineering, fiber technology, materials science, or chemistry, adding, "engineering backgrounds may include chemical, mechanical, agricultural, materials, fiber, or process engineering." The biomass industry development coordinator position calls for an undergraduate degree. But note the last line of the job listing:

> Minimally a Bachelor's degree is required in Business Administration, Economics, or Marketing Research, with a minor in biology, chemistry, engineering or environmental science. Legal studies and law degrees that include a science background may also be good preparation for a career in this field. Basic understanding of scientific concepts and terms is key. Master's and Ph.D. degrees required for advancement.

In short, many of the positions within the field of bioenergy require at least an undergraduate degree, with a marked preference shown for graduate work. One recent listing for a job with British Airways called for only a BA, but a quick look at its requirements suggests that graduate training would come in handy:

> Responsible for project management, challenging and influencing BA departments on environment policy and programs. Working without direct supervision, using initiative, making decisions and recommendations as appropriate to the team and management in other BA departments. Responsible for delivering key aspects of environment projects, including carbon management, assisting in the development of the BA position on sustainable biofuels and analysis to support the global climate policy program.
>
> Manage the development, coordination and implementation of a comprehensive carbon information system. Compile carbon footprint and carbon efficiency outputs.
>
> Managing a project to assess automated Carbon Reporting packages for the business, including contract and supplier management.
>
> Project management skills in support of sustainable biofuels

program, including sustainability, commercialization and policy aspects. Project support on global climate policy program, including analysis of options for carbon neutral growth and implications for market distortion.

Similarly, a job listed by the industry leader SG Biofuels, headquartered in San Diego, California, but with installations in several countries, calls for a project manager in Guatemala. This requires only a bachelor's degree—though note the suggestion that an MBA would likely win a tie.

SG Biofuels is currently seeking a Project Manager to join our team in Guatemala. SG Biofuels is an energy crop company developing and delivering high performance bioenergy solutions for the renewable fuel, biomass and chemicals markets.

Responsibilities: Reporting to the General Manager, the Project Manager will responsible for managing the operations including land search, soil preparation, planting, agronomic practices and harvest of the crop. Additional responsibilities include: Supervise a team of operations professionals; Set and achieve targeted production volumes by managing the production function; Oversee and test different agronomic practices and equipment and manage the

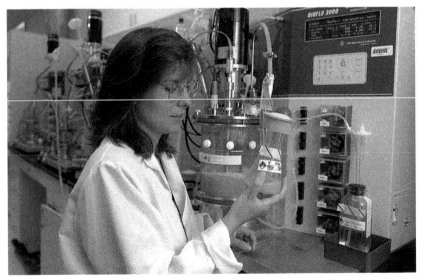

A researcher works on cellulose production fermentations to make enzyme for biomass-to-ethanol process. Photo: Dave Parsons, NREL

quality control of the process; Manage the product and logistics activities from harvest through delivery to the industrial process.

Requirements: Bachelor's degree in Plant Science, Agronomy, Horticulture, or Crop Science or a related discipline plus 5+ years of relevant experience in the management of big plantations is required; an MBA degree is preferred. Background with perennial crops is a strong plus (sugar cane, palm or bananas).

Excellent knowledge of administration of production, quality, and logistics technologies and processes, in addition to experience addressing seasonal challenges while meeting strict quality standards required.

Must have a successful track record in delivering a high production levels and within budget. Strong project management skills and ability to manage multiple tasks within distributed organization and work teams. Exceptional oral, written, presentation and interpersonal communication skills. Languages: fluent in both Spanish and English.

Fortunately, the field is broadly defined both within the realms of engineering and agriculture, programs that hundreds of colleges and universities offer. One intriguing one is a minor offered by Oregon State University, which can be combined with a major field of study such as engineering, chemistry, or economics to yield a very marketable set of knowledge and skills. The minor takes in the following courses, for a total of 28–34 credits:

- Introduction to Regional Bioenergy: Study Tour
- Interdisciplinary Research: Bioenergy focus
- Bioenergy and Environmental Impact
- Bioenergy Electives: Choose one course from each category (Technical; Environmental; Social/Economic/Policy)
- Research
- Thesis
- Projects - Data Presentation
- Seminar

And, as we noted earlier, there is just as much demand for trained technicians as there is for PhDs. Walla Walla Community College, for instance, offers an associate's degree, normally undertaken in two years, in bioenergy operations. That program is diverse by design, so that someone holding it can enter the bioenergy sector in any number of areas, most of them agricultural. Here are just two quarters' worth of classes for Year One:

Fall Quarter (17 credits)
- Agricultural Chemistry
- Introduction to Bioenergy
- Refrigeration and Air Conditioning Basics I
- Principles of Electricity Theory

Winter Quarter (16 credits)
- Biorefinery Processes
- Principles of Electricity AC Application
- Industrial Safety in the Workplace
- Introduction to Technical Mathematics

Ohio State University's Agricultural Technical Institute (*ati.osu.edu*) has developed program standards for a two-year associate's degree in renewable energy, with a specialty in bioenergy, which, to quote the ATI site, "focuses on the generation of biogas from organic material from agricultural, industrial, and municipal byproducts and waste." In a similar vein, many other programs are being created nationwide to cover technical education in bioenergy, including an increasing number of online options. As ATI's web page notes, it is expected that this training will open the door to such jobs as these:

A pile of biomass wood chips wait to be gasified at the McNeil Generating Station in Burlington, Vermont. Photo: Warren Gretz, NREL

- Equipment installation and service technicians
- Environmental engineering technicians
- Bioenergy plant/system operators
- Energy facility manager
- Renewable energy equipment sales
- Research laboratory technicians

See the listings at the back of this book for more programs, and narrow your search by using College Source Online (*www.collegesource.org*) or a similar web site. A good source of knowledge on trade jobs and community college training is from the aforementioned Energize Iowa (*http://energizeiowa.org/energy-green-job-training.html*).

Another useful source of information for jobs requiring varying levels of education and experience is the British sustainable-energy recruiting firm Allen & York (*www.allen-york.com/renewable-energy-jobs/worldwide/bioenergy-and-efw-jobs*). Note, of course, that most of those jobs are within Britain or the European Union and are typically not available to citizens of nations outside that community. Even so, the listings give a good idea of what kind of qualifications are required in order to win one of the many kinds of jobs within the bioenergy sector. ❖

A chemist conducts compositional analysis of agricultural residue (corn stover) for carbohydrate and lignin content. Photo: Warren Gretz, NREL

Resources

Organizations & Web Sites

Biomass Energy Businesses in the World
energy.sourceguides.com/businesses/byP/biomass/biomass.shtml
This web site lists bioenergy firms of various kinds and is a useful first stop in planning a job search.

Environmental and Energy Study Institute
www.eesi.org
Founded by a bipartisan congressional caucus in 1984, the Environmental and Energy Study Institute (EESI) is a 501(c)3 nonprofit organization dedicated to promoting environmentally sustainable societies. EESI advances innovative policy solutions that set us on a cleaner, more secure and sustainable energy path, including bioenergy.

Great Lakes Bioenergy Research Center
www.glbrc.org
GLBRC, a consortium of schools in Michigan, Ohio, Illinois, and Wisconsin, develops the basic science that will form the foundation of new biofuels technologies, turning today's energy challenges into energy opportunities for the future.

University of Illinois Center for Advanced Bioenergy Research
bioenergyuiuc.blogspot.com
This blog, produced by the Center for Advanced Bioenergy Research (CABER) at the University of Illinois, offers a useful roundup of research news and related topics dealing with biofuels. It is regularly updated.

U.S. Department of Agriculture; Careers with the Agricultural Research Service
www.ars.usda.gov/Careers/Careers.htm
This federal web site contains listings for a wide variety of positions within the Agricultural Research Service, where significant work in biofuels is being carried out.

U.S. Department of Energy
Energy Efficiency and Renewable Energy: Biomass Program
www1.eere.energy.gov/biomass/for_students.html
This web site offers abundant resources for students, including information on various biomass technologies and sites devoted to jobs and careers.

Publications

Biofuels Engineering Process Technology by Caye Drapcho, John Nghiem, and Terry Walker (McGraw-Hill Professional Publishing, 2008). Written by a team of experts from Clemson University with experts from around the world, this book is a comprehensive description and discussion of the concepts, systems and technology involved in the production of fuels on the industrial and individual scales.

Biodiesel Power by Lyle Estill, (New Society, 2005). Lightly touching on the technical aspects of the fuel, its qualities and specifications, this book is largely about the people and stories of the biodiesel movement.

Biodiesel Basics and Beyond: A Comprehensive Guide to Production And Use for the Home and Farm by William Kemp (Aztext Publishing, 2006). A useful handbook on how to make biodiesel fuel, including processes, lists of equipment, and a compendium of reference materials and suppliers.

Biofuels and Bioenergy: Processes and Technologies by Sunggyu Lee and Y.T. Shah (CRC Press, 2012). A comprehensive reference/textbook covering the methods used in developing biofuels and their applications.

Introduction to Biofuels by David M. Mousdale (CRC Press, 2010). This survey text covers key technologies of biofuel manufacturing, including biotechnology, bioprocessing, and genetic reprogramming of microorganisms.

Biofuels by Lisbeth Olsson, ed. (Springer, 2007). A Ph.D.-level series of technical papers on various topics in bioenergy, useful for giving a sense of present and future research emphases.

Biodiesel: Growing a New Energy Economy by Greg Pahl. (Chelsea Green, 2005). Written for a nontechnical readership, Pahl's book surveys the history of the biofuels industry, assessing successes and current shortcomings and offering insights into its future.

Biodiesel America by Josh Tickell (Yorkshire Press, 2006). Tickell shows that an abundance of available, economically viable, and profitable energy solutions exists, and at the forefront of these new energy technologies is biodiesel.

Fuel to Make the Wheels Turn

The U.S. uses more than 140 billion gallons of gasoline and almost 40 billion gallons of diesel fuel annually.

More than 60 percent of the petroleum we use is imported.

Source: Biomass Conversion Research Laboratory, Michigan State University

Where to Study

California Institute of Technology

www.caltech.edu
1200 East California Blvd.
Pasadena, CA 91125
(626) 395-6811

Caltech is a leading research institution in the study of renewable energy, especially solar energy and biofuels. Its chemistry department, to name just one program, sponsors research in solar energy conversion and storage, methane oxidation, and nitrogen fixation.

California Polytechnic State University

www.calpoly.edu
College of Agriculture, Food and Environmental Sciences
San Luis Obispo, CA 93407
(805) 756-2161

The BioResource and Agricultural Engineering major at Cal Poly offers hands-on experience in a wide range of engineering skill areas, renewable energy and waste treatment, electronics and control systems, and resource information systems. The BRAE program is accredited by the Accreditation Board for Engineering and Technology (ABET).

Drexel University

www.chemeng.drexel.edu
Chemical and Biological Engineering
3141 Chestnut Street
Philadelphia, PA 19104
(215) 895-2227

Development of technologies for harnessing energy from renewable resources and for reducing the environmental impact of worldwide resource consumption is now receiving major national and international attention. Chemical and Biological Engineering Department faculty at Drexel University are leading innovative research programs related to energy and the environment, including the production of biodiesel from natural and waste oils containing high free-fatty-acid contents, developing new materials for higher efficiency solar cells, deriving polymers from renewable fatty acid monomer, and modifying the composition and structure of fuel cell catalyst layers for better performance.

Florida A&M University

www.famu.edu
College of Agriculture & Food Sciences
Tallahassee, FL 32307
(850) 599-3000

Florida has the most abundant biomass resources in the United States. Florida A&M, a historically black college (now university), offers a range of programs in bioenergy education, holding that it can be the catalyst for a new suite of industries in the state.

Michigan State University

canr.msu.edu
College of Agriculture & Natural Resources
102 Agriculture Hall
East Lansing, MI 48824
(517) 355-0232

Michigan State, an old and well-regarded land-grant college, is the home of the Biomass Conversion Research Laboratory. Its agriculture college offers several graduate, undergraduate, and two-year certificate programs in fields that pertain to renewable energy, including biosystems and agricultural engineering, construction management, crop and soil sciences, forestry, and agricultural technology.

Montana State University

www.sfbs.montana.edu
Sustainable Food & Bioenergy Systems
Bozeman, MT 59717
(406) 994-5640

The Montana State University SFBS program offers interdisciplinary undergraduate degrees in four concentrations: the Sustainable Food Systems Option, Sustainable Crop Production Option, Agroecology Option, and Sustainable Livestock Production Option. Each option includes coursework in sustainable food and bioenergy production.

Oregon State University

agsci.oregonstate.edu/bioenergy
College of Agricultural Sciences
137 Strand Agriculture Hall
Oregon State University
Corvallis, OR 97331–7304
(541) 737-2999

Oregon State offers numerous programs in bioenergy, including the minor described in the text.

Pennsylvania State University

www.bioenergy.psu.edu
Penn State Biomass Energy Center
225 Agricultural Engineering Building
University Park, PA 16802
(814) 865-3722

Penn State's Biomass Energy Center acts as a clearinghouse of information and a coordinator of campus-wide activities related to bioenergy. Its associated faculty come from several departments, and exploring their listings will yield information on a variety of programs. In addition, the Center offers short courses for continuing education and professional development.

University of Colorado Environmental Center

ecenter.colorado.edu
University Memorial Center, Rm 355
Boulder, CO 80309-0024
(303) 492-8308

Colorado is one of the greenest campuses in the nation, literally and figuratively, and a leader in the use of renewable energy sources such as wind and solar power. The university offers a wide range of programs in every aspect of renewable energy. UC also houses the Colorado Center for Biorefining and Biofuels, which offers numerous short courses throughout the year.

Walla Walla Community College

www.wwcc.edu
Agriculture Center of Excellence
500 Tausick Way
Walla Walla, WA 99362
(509) 522-2500

The Bioenergy Department offers an associate's degree in bioenergy operations (see the text).

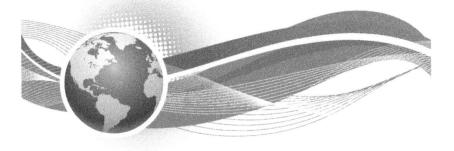

Green Building & Energy Management

The automobiles we drive and the airplanes that we fly use a lot of energy. The buildings in which we live, work, study, and shop are major consumers of energy, too: according to the U.S. Environmental Protection Agency, buildings account for 36 percent of total energy use, 12 percent of total water consumption, 65 percent of total electricity consumption, and 30 percent of the carbon dioxide emissions in the nation. As scholars, scientists, inventors, entrepreneurs, and workers of all kinds have turned their attention to improving the efficiency of vehicles, so they have increasingly studied how the built environment can be bettered. And so it can: Just five years ago, when the first edition of this book was published, the figures were respectively 39 percent, 12 percent, 68 percent, and 38 percent, respectively.

Look, for example, within your own home. Forget for a moment about the appliances and electricity-using things within it, energy-intensive though they are. Look instead at its veins and bones and skin and eyes, at its pipes and frame and covering and windows.

If there's a dripping faucet, it's dripping at a rate of as much as 20 gallons a day, a drop at a time.

If the house is drafty, then it can lose the same amount of heat or cooled air through leakage as it would if a window-sized hole were simply punched through a wall and left open to the elements. The loss from little leaks can amount to more than a third of your annual utility bill.

If the roof is old and in poor repair, then cool air is fleeing in summer and hot air is running away in winter. An old roof can add 10 or 15 percent to the base cost of your utility bill, and even more.

If the windows rattle, then they're leaking, too. Old windows will do that. A single-glazed, one-pane window has a thermal resistance value, or R-value, of 1: the window is almost the same temperature as the outdoors. Double-glazed windows, with an R-value of 2, are more efficient. "Superwindows," with panes separated by argon or krypton gas or some other insulating substance, have an R-value of between 6 and 10.

Energy management specialists and technicians, builders, and tradespeople have known these realities for a very long time, but it's only been in recent years that technologies have been available to take full advantage of a building's ability to provide shelter while using the fewest possible resources, especially nonrenewable ones. The quest to renovate, and rebuild, and build from scratch intelligently, economically, and sustainably falls under the informal rubric "green building."

There's green in more ways than just environmental friendliness and energy efficiency. Ten years ago, the phrase "green building" might have meant something to a handful of theorists and the visionaries of Berkeley and Cambridge. As of 2010, the green building industry was worth $12 billion, while about 10 percent of commercial construction starts were green by then—a number that had increased to nearly 20 percent just two years later. Additionally, foreign markets were a source of employment for American workers and specialists trained in LEED standards, LEED being what the U.S Green Building Council defines so: "LEED (Leadership in Energy and Environmental Design) is a voluntary, consensus-based, market-driven program that provides third-party verification of green buildings." For example, China and other relatively strong economies saw significant expansion in energy-efficient construction. Notes Rob Watson in the *Green Building Market and Impact Report 2011*, "LEED has registered or certified projects in 131 of the world's 196 countries, with a total floor area of almost 3-billion sq. ft. When combined with the nearly 1.5-billion sq. ft. registered under LEED India and LEED Canada, it is clear that LEED is the dominant global green building certification brand."

"If you're going to be part of the construction business for the next thirty or forty years, then you're going to have to make sustainability and green building a big part of your repertoire," says Gregory Esau, a builder and entrepreneur in Vancouver, Canada. "A lot of municipalities, for instance, are going carbon-neutral, and we're going to have challenges figuring out how to do that—not just in the buildings themselves, but also in building the buildings, what with all the big machines we have to use! We'll need good people to help us figure out solutions to problems like that, workers who are on what I call a green track."

Asked for advice he might give to a student with an interest in green building, Esau enthusiastically counsels, "Try to be a leader! Do your homework, learn as much as you can, ask a lot of questions. You can make a good living at this work, and it's a very exciting time to get into it."

Careers in Green Building

Green building, one source defines it, is "the practice of creating healthier and more resource-efficient models of construction, renovation, operation, maintenance and demolition." That's a tall order, particularly when it comes to the challenge of refitting old buildings greenly (so to speak) and taking down the ones that cannot.

A green building requires visionaries. It requires women and men who can study the land to determine where the sun will fall in winter and summer and where the wind will blow, the better to create efficient passive-solar systems for heating and cooling, the better to place energy-generating photovoltaic panels and wind turbines, the better to know where a geothermal pipe will find its happiest home. It requires architects, contractors, and builders who are committed to finding the most sustainable path to the best building they can make. It requires real-estate agents, lawyers, salespeople, accountants. It requires scientists who understand the physics of heat transfer, the flow of water, the movement of the winds. It requires engineers of every description. It requires men and women who can put those systems into place and maintain them: carpenters, plumbers, electricians, roofers, landscapers, glaziers and window installers.

The list goes on and on. Some of these careers are relatively new, while others are as old as humankind. Some are a hybrid. For example, glaziers and carpenters who work with new energy-efficient windows are combining skills collectively learned over hundreds of years with emergent technologies, while specialists in retrofitting older buildings necessarily have to work with structures that can be hundreds of years old while keeping a thought for the future.

All that said, things have quieted down for many in the construction trade since the so-called Great Recession hit in full force in 2008. While parts of the country have experienced a modest resurgence in the sales of new homes, and while major companies have slowly been investing more resources in the improvement of their physical plants, many other parts of the country and sectors of the economy have been feeling a prolonged squeeze. No one can predict how the economy will turn, especially while politicians play their power games with it, but it can be said with confidence that when builders are hired these days, it is very often because they have skills in these new technologies, from installing super-efficient glass windows and radiant-heated floors to taking homes off the grid with solar panels and geothermal exchange units.

In short, although the economy remains uncertain and hiring patterns still vary widely from region to region, there's room for just about anyone in the green building field—anyone, that is, who is willing to work hard, learn, and keep working hard and learning for all the years to come.

Knowledge is Power

Green building, at whatever level and whether approached from the white-collar, blue-collar, or green-collar perspective, requires specialized knowledge. Workers in the building trades are often expected to have that training in hand before going on the job, acquired from community-college or trade-school courses and certifications. One interesting sequence of courses, apparently geared to workers with on-the-job skills, is Mount Hood Community College's Sustainable Building Advisor Training Program (*www.mhcc.edu/sbap*).

For training in the trades in your area, do an Internet search combining the terms "green building" and "community college," along with your location. Doing so in the city in which I live turns up, for instance, a site from Pima Community College devoted to the construction trades, and it's well worth a look wherever you are: the site is at *www.pima.edu/programs-courses/credit-programs-degrees/index.html*.

Workshops such as those offered by Solar Energy International (*www.solarenergy.org*) and Yestermorrow (*www.yestermorrow.org*) can provide basic training and continuing education, and apprenticeships are opening up as more and more skilled workers arrive on the scene and are available to serve as mentors.

For most "green-collar" positions, a college degree is preferred. Here, for instance, is a listing for a position as a green building project analyst working in the Southeast:

Provide sustainable design support services for planning, implementing, and follow-up to ensure the facility's construction projects fully meet the goals of sustainable design. Participate in early construction project planning. Attend, coordinate, and facilitate (senior level) LEED® credit dialogue and negotiations during planning and design charrettes; and assist client staff, as needed, in managing the sustainable design portion of each project. Provide technical guidance/recommendations for potential LEED credits that may be achievable and practical. Provide assistance in adding credits to each project's sustainable design rating whenever feasible. Identify initial sustainable design credit goals and identify any changes to those goals as they occur and include the rationale for the changes. Inform management of any new sustainable design policies or requirements.

The job calls for specialized training—but, at the outset, a college degree:

Minimum Qualifications: Bachelor's degree (preferred Master's Degree) in architecture, civil or environmental engineering, environmental studies, environmental sciences, renewable energy, sustainability planning, natural resources management, or related field. Professional training and work experience in energy efficiency, sustainability planning, master planning and/or sustainable design and construction. Knowledge of, and experience implementing, the U.S. Green Building Council's LEED guidelines; and Federal environmental laws and regulations and pollution prevention programs. Good computer skills in MS Office software are essential. Must have strong organizational and interpersonal skills to successfully coordinate and facilitate sustainable design portions of planning and design charrettes, briefings, and related activities. The on-site personnel must have the ability to work well as a team member.

Thousands of colleges and universities offer programs that might lead to such a job. Some are well-established flagship schools such as MIT and Yale, whose architectural programs are second to none; others are more experimental. One was the now-defunct New College of California, whose North Bay Campus offered a master's degree in Eco-Dwelling; though the school is gone, there are many ways to appreciate what might be called the hippie building arts, one great source for which is Lloyd Kahn's Shelter Publications (*www.shelterpub.com*) and its founder's excellent blog (*lloyd-kahn-ongoing.blogspot.com*).

You stand to learn a lot wherever you choose to study, for green building is a field that's being defined and redefined every day, where experimentation and entrepreneurship are at a premium. To find the school that's right for you, ask yourself what kind of work you'd like to be doing ten years from now. If it's building large, self-contained structures and experimental communities such as Arcosanti, in the Arizona desert, then a blend of architecture and engineering such as MIT offers might be just right. If it's siting buildings and communities to take best advantage of the natural environment, then a degree in environmental studies such as the excellent one the University of Washington offers could be the key. Talk to architects, builders, and engineers in your community and take their advice. If you live in a good-sized town or city, the chances are good that the local government will have a planning department, and within such departments are often brilliant, well-educated thinkers and doers who have considerable expertise in green building and energy management—and a solid sense of where local job possibilities might lie.

Explore. A green track awaits.

. . . .

LEED Gold: The Solaire 27-Story Residential Tower

Consider the variety of skilled workers needed to design and construct this Gold-certified LEED building in New York: The Solaire—a 27-story residential tower with 293 units—cut its energy demand by 35 percent using automatic dimming fluorescent lights, high windows, daylighting and other strategies; west-facing photovoltaic panels supply 5 percent of the building's energy needs. Ninety-three percent of the construction waste for the project was recycled and about 60 percent of the building materials were made from recycled content. To maintain superior air quality, the building features filtered fresh air, operable windows and controlled humidity. Its residents have access to public transportation,

on-demand hybrid rental cars, bicycle parking and electric vehicle charging. Gardens of native shrubs, perennials and bamboo cover 75 percent of the roof, helping to lower heating and cooling loads and increase tenant satisfaction. To help reduce potable water demand by 50 percent overall, the building uses recycled wastewater for its cooling tower, low-flow toilets and for irrigating landscaping. *www.albaneseorg.com or www.thesolaire.com. Photo: Albanese Organization*

. . . .

Getting Into Green Building

Scott Sklar writes regularly for several online publications in the field. Here's his answer to an inquiry from a resident of upstate New York who asked about where a person interested in a career in designing residential and commercial projects might start:

I am receiving many e-mails from RenewableEnergyAccess.com readers on jobs in the industry, particularly in the design, sales, installation and service sector of the industries.

First, while the clean energy industries are growing more than 30 percent per year, the biggest bottleneck in this growth is in the product delivery chain to the customer. When I get asked, "How should I enter the field?" my first response is to approach existing service vendors.

In system design, architectural and engineering firms are a good start in your locality. They may have someone, either full or part time, assigned to the renewable or distributed energy or green building sectors, and they may want to add or train someone in providing services. So check with people locally or go to the local Yellow Pages.

Second, the existing solar integrator, design, sales, installations and maintenance firms are growing, and they may need to add employees. Contact them, again checking the Yellow Pages in your area, or even better at FindSolar.com or via the State Chapter lists from the websites of the American Solar Energy Society or the Solar Energy Industries Association.

I also urge approaching corollary industries that might want to set up some kind of parallel business in addition to their core business: HVAC contractors, building security companies, local telecommunications and cellular/tower companies, and traditional backup power companies now primarily offering diesel or battery banks—since they may want to expand to cleaner applications. (RenewableEnergyAccess.com lists job opportunities, which can be broken down to suit your search.)

Local unions, such as IBEW, have a concerted effort to train their members, and I am sure local electricians, carpenters and builders are also beginning to add jobs in this field. So think out what may be best in your area, reach out to established and new players.

Good luck in finding a job. We need lots of new blood in this field as we scale up and grow.

Taking Efficiency to a New Level

Michael Sykes invented the award-winning Enertia® Building System (U.S. Patent No. 6,933,016) in the mid-1980s *(www.enertia.com)*. As this conversation shows, Sykes affords a textbook example of how technological knowledge and entrepreneurship can lead to a highly successful career in renewable energy and green building.

Summer   Winter

To achieve super efficiencies, an Enertia home combines the effects of several principles, including the greenhouse effect, the thermal inertia (mass) of wood and earth, and the natural convective movement of heated air. The heavily windowed sunspace—a 4-foot-wide room running along the entire south side of the main level—captures winter heat and transfers it throughout the house via airflow channels between the ceiling and the roof, and the inner and outer north walls. In summer, cool air is pulled from the north side and exits through roof vents. It is, essentially, a solid wood house within an envelope of glass and wood. www.enertia.com

What motivated you to start building Enertia Homes? *Over the years I had made observations that struck me as out of the ordinary. Then one day they all came together, there was a light-bulb moment and I realized I had stumbled on a new form of energy. I made up a word for it…"Enertia." It would enable us to use nature's most abundant renewable material, wood, to build houses that heat and cool themselves without fuel or electricity. Amazingly, in 1984, no manufacturer or investor was interested so I reluctantly set about to make them myself. I could not let it go—here was a solution to a major worldwide problem. Finding investors and growing the business continues to be my biggest challenge.*

What education and skills do you use as an inventor of efficient green homes? *Enertia, energy from a change in phase, was so new that I had to write my own equations for it. I actually used the math I learned in high school* *and college, including differential equations. The air-flow patterns are based on physics and meteorology. The materials are based on biology and forestry, using nanotechnology and chemistry. The designs are based on architecture; the structure is based on engineering. To make the houses I had to learn industrial engineering and machinery design. Yet everything I learned as I developed my invention went against the conventional thinking in every field. When your invention is new you have to be an entrepreneur—by definition, no one else is doing it.*

. . . .

Careers in Energy Management

Buildings are major polluters, and they account for about a huge chunk of the energy we use in this country. On the positive side, though, they also afford an arena in which we can improve our habits of using energy, and building environmentally friendly structures, as we have seen, is one important path toward lessening our energy consumption overall. So is designing more efficient energy systems, putting them in place, and maintaining them.

The field of energy management is a broad one, and it requires familiarity not only with energy technologies but also with the building trades, making many jobs within it an interesting blend of blue- and green-collar work.

An energy engineer, for example, works with computer software to develop various scenarios of energy use for new building projects, then provides hands-on, quite specific advice to owners and contractors on how to improve energy efficiency. Many energy engineers come from the building trades, with backgrounds in HVAC, electrical systems, and general contracting; others have degrees in engineering or architecture and develop such skills by spending time apprenticing on construction sites and in other on-the-job situations.

An energy analyst has some knowledge of engineering, but spends most of his or her time on the job performing calculations and audits based on energy use and cost, usually after clients have requested help in identifying ways in which energy costs can be lowered in existing structures. Energy analysts are thoroughly familiar with building codes related to energy. Some analysts come from the building trades, although many energy analysts have training as electrical engineers. Certification is desirable, which requires passing a standards exam and attending yearly training classes and workshops.

HVAC technicians are also essential to the practical work of maintaining energy-efficient heating and cooling systems. In the past, most such technicians came up through the building trades, but because those systems are now so complex, today most come into the field following an apprenticeship or, more often, technical or trade school, with training in electronics and basic math emphasized.

An emerging option is in what Lane Community College calls an "energy management technician." Its two-year program is both rigorous and comprehensive, and it gives a good idea of what an interested student might look for in other programs around the country. As the LCC catalog describes it, the graduate will:

- Evaluate the energy use patterns for residential and commercial buildings and recommend energy efficiency and alternative energy solutions for high-energy consuming buildings.
- Understand the interaction between energy-consuming building systems and make recommendations.
- Construct energy evaluation technical reports and make presentations for potential project implementation.
- Use appropriate library and information resources to research professional issues and support lifelong learning.

The Energy Systems Integration Laboratory being built at NREL. Photo: Dennis Schroeder, NREL

- Access library, computing and communications services; obtain information and data from regional, national and international networks.
- Collect and display data as lists, tables and plots using appropriate technology (e.g., graphing calculators, computer software).
- Develop / evaluate inferences and predictions based on that data.
- Determine an appropriate scale for representing an object in a scale drawing.
- Interpret the concepts of a problem-solving task, and translate them into mathematics.

That's quite an ambitious program, and it gets even more ambitious with the addition of the Renewable Energy Technician Option, the graduate of which, LCC adds, will:

- Appropriately size and recommend renewable energy system types for particular situations.
- Understand and put into practice the installation protocol for Photovoltaic and Solar Domestic Hot Water Systems.

How to get there? Well, here's LCC's list of required courses:

First Year
Introduction to Spreadsheets and Databases
Blueprint Reading: Residential and Commercial
Intermediate Algebra
Introduction to Energy Management
Introduction to Sustainability
Residential/Light Commercial Energy Analysis
Alternative Energy Technologies
Introduction to Water Resources
Co-op Ed: Energy Management Seminar
Fundamentals of Physics
English Composition
Air Conditioning Systems Analysis
Energy Efficient Methods
Lighting Fundamentals
Fundamentals of Physics
Human Relations at Work

Second Year
Commercial Air Conditioning Systems Analysis
Lighting Applications
Energy Investment Analysis
Technical Writing
Commercial Energy Use Analysis
Energy Control Strategies
Co-op Ed: Energy Management Seminar
Building Energy Simulations
Energy Accounting
Co-op Ed: Energy Management

Add to this program other requirements and a few electives, and you'll see that Lane Community College students are busy—but also very well prepared to enter the workforce at levels commanding initial pay of at least $30,000 a year. In fact, LCC reports that the starting pay for its graduates are $38,000–45,000 for energy management specialists, $25,000–35,000 for a renewable energy technician, and $40,000–50,000 for a specialist in resource conservation management. ❖

Saving Energy

Energy-efficiency and renewable-energy author-ity Amory Lovins, the head of the Rocky Mountain Institute, has been working to develop "soft-path" solutions to the energy problem since the 1970s. He believes that one key to saving energy is to put better lighting systems into use: a restaurant, he calculates by way of example, can cut 80 percent of its electricity costs simply by installing energy-efficient light bulbs. "There's upwards of a hundred giant power plants to be saved by proper lighting sys-tems," he told writer Elizabeth Kolbert in a *New Yorker* profile published in January 2007. Another smart solution is to replace old windows with new double-paned, gas-sealed ones; in one Chicago office building, he determined, the savings from that retrofitting—an expensive process—would be so quickly realized that payback would come "in minus five months." Making calculations of this sort is the job of energy gurus—and energy managers.

Managing Energy Efficiently: Texas Instruments

In 2004, Texas Instruments embarked on an ambitious project to build the world's first "green," LEED-certified (Leadership in Energy and Environmental Design) semiconductor manufacturing facility in an effort to reduce construction and operating costs and the company's impact on the environment. The new fabrication site (RAFB) was located in Richardson, Texas.

Although building "green" required some additional investment to realize long-term operating benefits, it added up to less than one percent of the construction budget. In addition, the plant was successfully built for an estimated 30 percent less in cost than a similar TI manufacturing plant constructed just 6 miles away only a few years earlier. This latter achievement increased the building's cost competitiveness among other semiconductor manufacturing facilities being built outside of the U.S.

Before any design funding was approved for the construction of the facility, a small group of Texas Instrument employees began investigating sustainable design. They gathered information, compiled data and brainstormed ideas. As the research team began to understand what was possible in their drive toward sustainable design, they knew they needed to solicit management support. A research team member offered TI's senior vice president of Manufacturing a tour of his passive/active solar house. While the tour provided a good primer on sustainable design, it was the low operating cost that really caught the executive's attention. He wanted to know first and foremost, "How much of this design process scales up to a large facility?" The answer: "All of it!" The conversation ended with one last question, "What do you need to make this happen?"

About a month before design funds were approved, more than thirty TIers convened with a dozen folks brought in by Amory Lovins and the Rocky Mountain Institute (RMI). The team held a 3-day design charrette to brainstorm ideas, then analyze and prioritize them. This list was dubbed the "Big Honkin' Ideas." In the end, most of the Big Honkin' Ideas were incorporated along with dozens of other items that came from this meeting.

In addition to sustainable features of the 92-acre site and the utility building, the office building offers four distinct benefits:

Energy Savings

- Passive solar orientation minimizes unwanted sunshine and exterior shade screen minimizes summer heat.

- Light shelves reduce the need for indoor lighting by bouncing daylight deeper indoors.
- Reflective roofing reduces the urban heat island effect.
- Quality window glazing provides a balance of good insulation and good visible light transmission.
- Smart lighting has built-in motion and photo sensors to respond to indoor conditions.
- Solar water heating.
- Water turbine-powered hand wash faucets.

Water Savings

- Waterless urinals save 40,000 gallons of water each per year.

Improved Air Quality

- CO_2 sensor-controlled ventilation provides the intake of fresh air as needed.
- Use of safer building materials, including paints, sealants and adhesives.
- Locally manufactured materials reduced shipping pollution.
- Shuttle buses, free annual mass transit passes, and a carpool matching program will discourage single occupant commuting.

Reduced Material Use

- The recycled content of all materials used in building construction is greater than 20 percent:
- Ceiling tiles have a recycled material content greater than 80 percent.
- The carpet is made from recycled materials and has very low emissions.
- Recycling centers for employees.
- Use of certified wood and wheatboard ensured the wood was extracted in a sustainable manner and contributes to the preservation of old growth forests.
- High-velocity hand dryers conserve paper towels.
  Source: Texas Instruments

Resources

Green Building Organizations & Web Sites

BuildingGreen

www.buildinggreen.com
This company provides accurate and timely information to building-industry professionals and policy makers to improve the environmental performance and reduce the adverse impacts of buildings. Also publishes Environmental Building News.

Florida Solar Energy Center

www.fsec.ucf.edu
The FSEC building science program researches and develops building improvement strategies that reduce energy use, enhance the economy, and improve the environment. Research projects include Zero Energy buildings, fenestration, energy efficient schools and green standards.

Green Home Building

www.greenhomebuilding.com
A useful web site devoted to sustainable architecture, solar home heating, and hybrid architecture; the listings of workshops alone make the site worth bookmarking.

National Home Builders Association

www.nahb.org
Check out the Green Building, Remodeling & Development section for a wealth of information, including the National Green Building Standard.

Solar Energy International

www.solarenergy.org
For workshops on sustainable home design and natural house building, check their website for the latest schedule.

The Florida Solar Energy Center uses solar hot water, solar electricity, low-e windows, and light "shelves" for reflected light, just to name a few of its energy efficient features. Photo: Nicholas Waters, FSEC

Sustainable Buildings Industry Council

www.nibs.org
SBIC was founded in 1980 as the Passive Solar Industries Council. While it continues to be an association of associations committed to high-performance design and construction, its new name reflects its efforts in the fields of architecture, engineering, building systems and materials, product manufacturing, energy analysis, and "whole building" design. SBIC is a branch of the National Institute of Building Sciences, an important clearinghouse for information on green building and energy management.

U.S. Environmental Protection Agency

www.epa.gov/oaintrnt/projects
The EPA web site provides resources and publications for the study of green building, as well as listings of grants and other funding opportunities for green-building projects.

U.S. Green Building Council

www.usgbc.org
With more than 11,000 member organizations and a network of 75 regional chapters, USGBC is a nonprofit organization composed of leaders from the building industry working to promote buildings that are environmentally responsible, profitable and healthy places to live and work. Its web site is a mine of information and resources, including LEED (Leadership in Energy and Environmental Design).

Yestermorrow Design/Build School

www.yestermorrow.org
They offer over 100 hands-on courses per year in design, construction, woodworking, and architectural craft and offer a variety of courses concentrating in sustainable design.

Energy Management Organizations & Web Sites

American Society of Heating, Refrigeration and Air Conditioning Engineers

www.ASHRAE.org
ASHRAE, founded in 1894, is an international organization of 50,000 persons. It fulfills its mission of advancing heating, ventilation, air conditioning and refrigeration to serve humanity and promote a sustainable world through research, standards writing, publishing, and continuing education. The organization's web site carries information on all these matters.

Department of Energy, Federal Energy Management Program

www.eere.energy.gov/femp
The U.S. government is the largest single energy consumer in the country. The Department of Energy's Federal Energy Management Program (FEMP) works to reduce the cost and environmental impact of the Federal government by advancing energy efficiency and water conservation, promoting the use of distributed and renewable energy, and improving utility management decisions at federal sites.

Energy & Environmental Building Association

www.eeba.org
EEBA provides education and resources to transform the residential design and construction industry to profitably deliver energy efficient and environmentally responsible buildings and communities. Their comprehensive

web site offers numerous resources, including the Builders Guides that cover four different climates.

Energy Management Institute

www.energyinstitution.org

EMI offers specialized training in management and other aspects of the energy sector. Most of its work is with professionals in petroleum and other fossil fuels, but it has some experience with renewables. Its web site lists courses, workshops, and other professional-development tools.

Energy Star Program

www.energystar.gov

Energy Star is a joint program of the U.S. Environmental Protection Agency and the U.S. Department of Energy devoted to energy-efficient products and practices. The program certifies appliances that meet strict energy-efficiency guidelines. Its web site offers information and resources to help plan and undertake projects to reduce energy bills and improve home comfort.

Publications

Shelter by Lloyd Kahn and Bob Easton (Shelter Publications, 2000). A cousin of the venerable *Handmade Houses* and its kin in the hippie era, *Shelter*, published in its first edition in 1973, offers an inspirational how-to resource with more than 1,250 illustrations. The authors recount personal stories about alternative dwellings that illustrate sensible solutions to problems associated with using materials found in the environment—with fascinating, often surprising results.

More Straw Bale Building: A Complete Guide To Designing And Building With Straw by Chris Magwood, Peter Mack, and Tina Theirren (New Society Publishers, 2005). A practical book that covers the entire process of building a bale structure: developing sound building plans; roofing; electrical, plumbing, and heating systems; building code compliance; and special concerns for builders in northern climates.

Design with Nature by Ian McHarg (Wiley, 1995). Blending philosophy and science, McHarg shows how humans can copy nature's examples to design and build better structures.

Building Green: A Complete How-To Guide to Alternative Building Methods: Earth Plaster, Straw Bale, Cordwood, Cob, Living Roofs by Clark Snell and Tom Callahan (revised edition, Lark Books, 2005). One of the most venerable of green building books, with a broad range.

Green Building Guidelines: Meeting the Demand for Low-Energy, Resource-Efficient Homes by Sustainable Buildings Industries Council (4th edition, 2007). A builder-friendly manual

What's an EMCS?

Energy Management Control Systems are tools that promote energy efficiency by accurately monitoring energy consumption and building operations

that, though in need of refreshing, is applicable to homebuilders interested in exploring the notion of rethinking some of the design issues in their current product.

Green Building Products: The GreenSpec Guide to Residential Building Materials by Alex Wilson and Mark Piepkorn (3rd edition, New Society Publishers, 2008). You'll find descriptions and manufacturer contact information for more than 1,400 environmentally preferable products and materials in this book. All phases of residential construction, from site-work to flooring to renewable energy, are covered.

Your Green Home by Alex Wilson (New Society Publishers, 2006). Written for homeowners planning a new home, this book sets out to answer some of the big-picture questions relating to having a home designed and built.

Energy Management and Conservation by Clive Beggs (Butterworth-Heinemann, 2002). This textbook surveys energy-efficient building practices and issues of cost, supply, taxation and policy. Highly useful for advanced students.

Guide to Energy Management, 7th edition by Barney L. Capehart, Wayne C. Turner, and William J. Kennedy (CRC, 2011). An essential reference for energy managers with chapters written by leading energy professionals on HVAC optimization, improving operations and efficiency ratings of equipment, understanding power factors and energy usage and bills, and the like.

The Homeowner's Handbook to Energy Efficiency: A Guide to Big and Small Improvements by John Krigger and Chris Dorsi (Saturn Resource Management, 2008). The authors guide non-specialist readers through the process of assessing current energy usage and predicting the benefits and estimating the costs of remodeling options. With projects ranging from simple fixes to large-scale renovations, this book offers solutions for the energy-conscious homeowner, regardless of budget, technical ability, or time.

Renovating Old Houses by George Nash (Taunton Press, 2003). A thorough survey of what it takes to make an old house new again, with an eye to energy efficiency and utility.

A Zero Energy Home in Lakeland, Florida, with solar panels on the roof. Photo: Steven C. Spencer, FSEC

Optimizing Energy Efficiencies in Industry by G. G. Rajan (McGraw-Hill Professional, 2002). A practical textbook full of examples that are useful to students of energy engineering and analysis.

The Home Energy Diet by Paul Scheckel (New Society Publishers, 2005). Learn about measuring, metering, investigating, and considering habits related to household energy use, then how to quantify energy consumption and cost, and how to make informed decisions about cost-effective improvements and upgrades. The book explores the misunderstood concept of efficiency versus cost by comparing fuel costs and equipment choices, including the possibility of using renewable energy for meeting home energy needs.

Handbook of Energy Audits, 9th edition by Albert Thumann and William J. Younger (Fairmont Press, 2012). A comprehensive and practical reference on energy auditing in buildings and industry, containing all the information needed to establish an energy audit program. Accounting procedures, electrical, mechanical, building and process systems analysis, life-cycle costing, and maintenance management are covered in detail.

Energy Efficiency Manual by Donald R. Wulfinghoff (Energy Institute, 2000). If it has to do with energy conservation, this massive (and expensive) handbook has something to say about it. Wulfinghoff, an energy expert, provides scenarios for 400 energy-conservation actions set in homes, offices, commercial buildings, and industrial plants. *Publishers Weekly* says that it "answers just about any question [from] homeowner, plant manager, energy policy guru . . . as practically useful as it is informative."

A Building America staff member tests a newly built home for energy efficiency by conducting energy monitoring, at the furnace and at the electrical meter. Photos: Warren Gretz, NREL

Where to Study

Arizona State University
schoolofsustainability.asu.edu
School of Sustainability
PO Box 87511
Tempe, AZ 85287–2511
(480) 727-6963

The recently established School of Sustainability at Arizona State University offers courses in sustainable environmental resources, resource allocation, energy and material use, engineering, and many other subjects. ASU's College of Technical and Applied Science also offers courses in renewable and alternate energy sources, including fuel cells (see technology.poly.asu.edu).

Fitchburg State University
www.fitchburgstate.edu
160 Pearl Street
Fitchburg MA 01420–2697
(978) 665-3000

Fitchburg State offers a range of courses in energy-related fields, including programs in industrial technology, architectural technology, energy management technology, and facilities management. The architectural technology concentration offers the study of technical systems in architecture, including energy systems, and ends with the development of professional practices.

Lane Community College
www.lanecc.edu/sustainability/energy-management-program
4000 East 30th Avenue
Eugene, OR 97405
(541) 463-3000

See the section on Energy Management for a description of LCC's extensive programs.

Syracuse University
www.syr.edu
Syracuse, NY 13244
(315) 443-1870

The Building Energy and Environmental Systems Laboratory (beesl.syr.edu) in the Department of Mechanical and Aerospace Engineering is a key research lab. Its mission, in part, is to advance the science and develop innovative technologies in the areas of indoor environmental quality and building energy efficiency. BEESL's labs are used for studies of combined air, heat, moisture and contaminant transport through building envelopes; interactions between indoor, outdoor environments and HVAC systems/components; sensitivity, accuracy and reliability of environmental sensors and control systems; and many other topics of interest to energy managers and students of renewable energy.

University of Alaska Southeast
uas.alaska.edu
11120 Glacier Hwy.
Juneau, AK 99801
(907) 796-6000

The UAS offers a program that teaches the essentials of building diagnostic assessment and building durability, performance, and energy efficiency. One-year certificate programs are offered in residential/light construction, building energy retrofit technology, and power technology, among other energy-related fields.

University of California at Los Angeles

www.ucla.edu
Department of Architecture
Los Angeles, CA 90095
(310) 454-7348

The internationally recognized UCLA architecture program offers a specialization in energy design and climate-responsive design. Credit is given for AIA continuing education; for more information, see: *www.aud.ucla.edu/energy-design-tools*.

University of Florida

www.floridaenergy.ufl.edu
Florida Energy Extension Service
P.O. Box 118300
Gainesville, FL 32611-8300
(352) 392-0947

The University of Florida provides continuing education courses in green building for Florida licensed building contractors and architects, as well as other interested parties. The extension service involves other state universities and a network of institutions elsewhere in the South—and even a partnership with Yale University.

University of Illinois at Chicago

www.erc.uic.edu/cap/engsolutions.htm
Energy Resources Center
851 S. Morgan Street
Chicago, IL 60607–7054
(312) 996-4490

The Energy Resources Center is an interdisciplinary program whose areas include energy management assessments, economic modeling, analysis of policy and regulatory initiatives, and education. ERC provides energy auditor/energy efficiency training to a selected team of high-school students each year, as well as an energy camp that offers a week of more intense training to the same students.

University of Massachusetts

www.ceere.org
Center for Energy Efficiency and Renewable Energy
160 Governors Drive
Amherst, MA 01003
(413) 545-4359

The Center for Energy Efficiency and Renewable Energy (CEERE) provides technological and economic solutions to environmental problems resulting from energy production, industrial, manufacturing, and commercial activities, and land-use practices. CEERE offers research, training and educational experiences for graduate and undergraduate engineers and scientists.

Washington State University

www.energy.wsu.edu
Energy Program at Olympia
905 Plum St. SE
Olympia, WA 98504-3165
(360) 956-2000

The Energy Program teaches energy and ventilation codes for construction staff and designers. The WSU web site offers extensive information on technical and trades-related issues.

Yavapai College

yc.edu
Construction Technology Department
2275 Old Home Manor Way
Chino Valley, AZ 86323
(928) 717-7720

Students learn through hands-on design and construction of a high-performance house using renewable energy. The college offers a one-year advanced certificate or a two-year AAS degree in residential building technology.

Hydrogen Energy & Fuel Cells

Hydrogen, one of the building blocks of the universe, is abundant. It also burns very cleanly, with the sole byproduct of burning being— well, water, for which reason talk of a "hydrogen economy" has been with us for many years. That economy is some distance off: hydrogen is difficult to transport, since it is highly flammable and requires considerable pressurization, and mass-production facilities have yet to be built. Still, a hydrogen-based fuel cell is around 60 percent efficient at converting fuel to power, double the efficiency of an internal combustion engine, and efficient compared to many other kinds of renewable energy storage and delivery systems. At some point in the not too distant future, Hydrogen

Buses are just one of the many places that fuel cells are used. Photo: Hydrogenics.com

is going to play an important part in the world's energy mix, and making that happen will open up many opportunities for scientists, technicians, and other workers.

One man who has been working in the field for many years now is Giorgio Zoia, who saw his first hydrogen-powered car at an exhibit in his native Italy. He was so impressed by it that he set out a rigorous course of education in science and mechanics, and for many years he has worked in the renewable-energy division at British Petrol (BP). In the first decade of the 21st century, he collaborated with DaimlerChrysler to maintain a fleet of 30 of the approximately 100 fuel-cell cars in California, where BP has its research facility. Each of those cars cost a million dollars. But, Zoia said at the time, their price would fall as hydrogen production and fueling stations come online—perhaps as soon as 2012.

That has yet to happen, almost certainly because of the recent eco-

5 Main Types of Fuel Cells

◆ alkaline
◆ molten carbonate
◆ phosphoric acid
◆ proton exchange membrane
◆ solid oxide

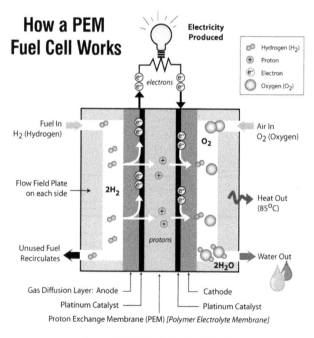

Source: HYDROGEN—Hot Stuff, Cool Science book

nomic downturn, which has delayed so much of the renewable energy economy in so much of the world. Even so, research in hydrogen power continues. One promising area of research in recent years, for instance, has been the use of photochemical molecular devices to produce hydrogen gas from water. A few years ago, chemists from Virginia Tech—a world leader in hydrogen technology, as well as renewable energy generally—unveiled devices called supramolecular ruthenium(II), rhodium(III) mixed metal complexes that use the energy from light to collect electrons, which are then used to split the hydrogen and oxygen in water. The researchers have been striving to make the process more efficient and in the past year, they have come up with additional molecular assemblies that absorb light more efficiently and activate conversion more efficiently. "We have come up with other systems to convert light energy to hydrogen," said Professor Karen Brewer. "So we have a better understanding of what parts and properties are key to having a molecular system work. Previously we concentrated on collecting light and delivering it to the catalyst site. Now we are concentrating on using this activated catalyst to convert water to hydrogen. Once we know more about how this process happens, using our supramolecular design process, we can plug in different pieces to make it function better."

An existing technology that has been widely employed in Britain is the combined heat and power (CHP) system, which produces both electricity and usable heat energy. One, used in the town hall of Woking, England, has a fuel cell that uses phosphoric acid as the electrolyte, natural

Ways to Liberate Hydrogen

Electrolysis split water with electricity

Steam Electrolysis split water with heat, pressure and electricity

Direct Solar Thermal Water Splitting split water with heat

Photoelectrochemical split water using sunlight directly

Thermochemical split water using chemicals and heat

Biological split water using microorganisms

Steam Reforming convert the methane in natural gas using steam

Direct Thermal Splitting of Natural Gas split natural gas using heat

Gasification breakdown coal or biomass with heat and pressure

Source: *HYDROGEN—Hot Stuff, Cool Science* book

gas reformed into hydrogen as the fuel, and oxygen taken directly from the air as the oxidant. The first installed fuel cell was able to supply 200kW of electrical power, with the idea that it would operate alongside other renewable energy sources such as solar power and wind power to provide a greater capacity for the community in the coming years.

In Britain as well, bioscientists at the University of Birmingham demonstrated a reactor that used hydrogen-producing bacteria and sugar waste to generate electricity via a fuel cell. Interestingly, the reactor used waste products such as diluted nougat and caramel waste from the candy industry as food for the bacteria, waste that usually ends up in landfills. Once in the reactor, the bacteria consumed the sugar, producing hydrogen and organic acids. A second type of bacteria was then introduced into a second reactor to convert the organic acids into more hydrogen. The hydrogen was then fed to a fuel cell, in which it reacted with oxygen in the air to generate electricity. In the bargain, waste biomass left behind by the process was removed, coated with palladium, and used as a catalyst in another process to remove pollutants such as chromium and polychlorinated biphenyls (PCBs) from the environment.

If some of this seems the stuff of mad-scientist labs, that's to be expected. Renewable energy writ large is forward-looking, but hydrogen power verges on the futuristic—and thus near the realm of science fiction—in its potential to alter our energy regime.

Honda solar-powered water electrolyzing hydrogen station. Photo: Honda

Careers in Hydrogen Energy and Fuel Cells

Although the hydrogen economy lies in the future, many opportunities exist today. Many are in the research field, largely within academia or the government, while other technical positions are available at private firms.

One solid-oxide fuel-cell (SOFC) manufacturer, for instance, advertised for a manufacturing equipment engineer. The job required a bachelor's degree in mechanical, electrical, or ceramic engineering or equivalent experience; two to eight years of related field experience; general knowledge of manufacturing processes; experience with structured problem-solving processes and process-capability techniques; knowledge

A technician builds a Fuel Cell Power Module in the Hydrogenics plant. Photo: Hydrogenics.com

of PFMEA (process failure modes and effects analysis), equipment design, ergonomics, and safety standards as related to manufacturing processes.

Another manufacturer posted a listing for a process manager in the company's research and development division. The specific job responsibilities included:

- Developing innovative and enabling SOFC materials, process, and component technologies.
- Writing successful research proposals to private and governmental research sources.
- Establishing and directing experimental research plans for multi-year research programs.
- Contributing to the company's intellectual property portfolio.
- Assuring the transition of technologies from laboratory to commercial products.
- Managing costs and resource utilization while achieving project goals.

- Producing comprehensive reports detailing technical progress and plans.
- Presenting research results at technical conferences.

Critical qualifications, the company added, included a PhD in materials science, chemical engineering, or a related discipline, and a minimum of five years of technology development experience in fuel cells or other high-temperature ceramic electrochemical systems. "The ideal candidate will have a comprehensive understanding of the state-of-the-art in fuel cell technology, a worldwide reputation for technical achievement and innovation, and a proven track record for solving problems and achieving technical success," the listing concluded. "Prior technology commercialization experience is of particular interest."

In late August 2013, General Motors advertised several openings for fuel-cell engineers and designers. One such posting follows:

Designer Fuel Cell-ENG0016748 Initiate, analyze, interpret and express specification data necessary for the procuring, scheduling, manufacturing, merchandising, assembling, servicing and reporting of fuel cell systems and their components. Work involves numerous technical duties and decisions made within the limits of general practices, standards and policies. Successful candidates must have extensive knowledge of Vehicle and Powertrain configurations.

Key Responsibilities: Interprets design information and converts into model/option or assembly usage statements; Develops and maintains fuel cell system description summaries; Initiates, tests and coordinates specification computer system; Communicates engineering data and maintains specification liaison with other departments or units; Establishes a course of action to accomplish completion of the job and/or project; Coordinates input from internal/external customers to better understand customer needs and/or perceptions; Stays informed of current trends in areas of expertise; Actively identifies new areas for learning and takes advantage of learning opportunities.

Required Qualifications: Associate's Degree or equivalent experience; Ability to work independently and with others; Load/Maintain RDV structure using TcAE; Extensive knowledge of NX or similar 3D CAD systems (including advanced abilities); Self-starter;

Highly Motivated; Able to mentor Designers in regards to powertrain and vehicle data.

Preferred Qualifications: Understanding of basic mechanical concepts; Understanding of fuel cell systems and related component designs; Experience with engine and powertrain integration and packaging within vehicles; High level of analytical ability where problems are somewhat complex; Good oral and written communication skills; Knowledge of General Motors product policy and federal, state and local regulations; Basic understanding of data processing concepts; Demonstrated technical and professional skills in job-related area required; Appropriate interpersonal styles and communication methods to work effectively with business partners to meet mutual goals required; Working knowledge of TcAE (inc RDV and variants); Working knowledge of GPDS; Microsoft Office including Excel; Experience with Vehicle and Powertrain time lines (GM).

As you can see, qualifying for such a position requires a solid education, as well as experience. For more on careers in hydrogen energy and fuel cells, visit Fuel Cells 2000 (*www.fuelcells.org*). The work is there—so hit the books!

The U.S. Navy has used fuel cells in submarines since the 1980s, and alkaline fuel cells have flown over 100 missions and operated for over 80,000 hours in spacecraft operated by NASA. Photo: NASA

Knowledge Is Power

Creating the hydrogen economy will require the skills of many researchers and technicians, skills that are acquired almost entirely in universities, technical schools, and laboratories. Fortunately, the federal government and other agencies have been generous in funding scholarly research, and almost every state has at least one center for hydrogen fuel-cell studies. Some

An NREL researcher works on the compressor of the Wind to Hydrogen Project (Wind2H2). Photo: Ben Kroposki, NREL

of the most prominent or active are listed below; for a complete list, visit the U.S. Department of Energy's roster of schools now engaged in appropriate research and training (*www1.eere.energy.gov/hydrogenandfuelcells/education*). ❖

· ·

Focus on Fuel Cell Education

Fuel cells involve many branches of science and many technologies, and understanding them therefore requires solid grounding in several disciplines. At the Colorado Fuel Cell Center, a division of the renowned Colorado School of Mines, students must master the basics of chemical engineering, electrical engineering, mechanical engineering, and materials science. A specialized course follows in fuel-cell science and technology, introducing students to the fundamental aspects of fuel-cell systems, with emphasis placed on proton exchange membrane (PEM) and solid oxide fuel cells (SOFC). Students go on to learn the basic principles of electrochemical energy conversion, relevant topics in materials science, thermodynamics, fluid mechanics, electrochemistry, fuel-cell operation, system integration, and fuel processing. That's a curriculum!

· ·

Resources

Organizations & Web Sites

Florida Solar Energy Center

www.fsec.ucf.edu

(321) 638-1000

In 1997, FSEC was designated a Center of Excellence in Hydrogen Research and Education by the U.S. Department of Energy. The development of applications and industries that use FSEC's hydrogen and fuel cell research results has been a primary goal.

Fuel Cell & Hydrogen Energy Association

www.fchea.org

An industry association whose members include the world's leading fuel cell developers, manufacturers, suppliers and customers. Their site provides an excellent list of books, studies and reports, online publications, plus access to their monthly Fuel Cell Connection newsletter, and quarterly Fuel Cell Catalyst newsletter.

Fuel Cell Today

www.fuelcelltoday.com

Offers market-based intelligence on the fuel cell industry as well as an online career section, categorized by Academic, commercial & sales, Engineering, Managerial, Research, and Other.

Fuel Cells 2000

www.fuelcells.org

For more information on technical training, internships and more, visit their Career section.

International Association for Hydrogen Energy

www.iahe.org

The goal of this IAHE is to stimulate the exchange of information in the hydrogen energy field through its publications and sponsorship of int'l workshops, short courses and conferences, and they endeavor to inform the general public about the role of H2 energy. Their official scientific journal, International Journal of Hydrogen Energy, is published monthly. The web site provides recent international news, numerous links, conference details, and more.

National Fuel Cell Research Center

www.nfcrc.uci.edu

An excellent site for fuel cell information, activities, tutorials, FAQs, info about the University of California, Irvine's academic program, plus upcoming events, industry publications, educational resources and much more.

National Hydrogen Association

www.hydrogenassociation.org

Since 1989, NHA has provided data and educational materials to the media, safety and codes & standards officials, policy-makers, and the general public. They have over 100 members, including major industry, small business, government and university organization. Read the latest hydrogen headlines online, and check out their Hydrogen and Fuel Cell Job Board.

National Renewable Energy Laboratory

www.nrel.gov/hydrogen

Their web site provides a wealth of information about hydrogen and fuel cells, including their R&D efforts on H2 production and delivery, storage, fuel cells, technology validation, safety, codes and standards, and analysis. Learn how research activities crosscut and contribute to advances across the laboratory—in photovoltaics, bioenergy, transportation, wind, buildings, and basic sciences.

Publications

Designing and Building Fuel Cells by Colleen Spiegel (McGraw-Hill, 2007).

Fuel Cell Fundamentals by Ryan O'Hayre, Suk-Won Cha, Whitney Colella and Fritz B. Prinz (Wiley, 2009).

Fuel Cell Magazine is a bimonthly publication that serves managers and technical professionals involved in developing and applying fuel cell technologies worldwide. Subscriptions are free to qualified recipients. *www.fuelcell-magazine.com*

Fuel Cell Technology Handbook by Gregor Hoogers (CRC, 2002). A comprehensive book about the principles of fuel cell technology, plus various fuel cell applications and an expert's look at future developments.

Hydrogen & Fuel Cell Letter. Since 1986 editor Peter Hoffman has covered the science, business, economics, and politics of hydrogen and fuel cells—nationally and internationally. Published monthly. *www.hfcletter.com*

Hydrogen—Hot Stuff Cool Science 2nd Edition: Discover the Future of Energy by Rex A. Ewing (PixyJack Press, 2006). An enjoyable read for all ages, and a super resource for anyone looking to sort out the basics of hydrogen science.

Fossil Fuel Free Economy

Iceland has started to convert its fishing fleet from diesel engines to hydrogen fuel cells as part of a national project to create a fossil-fuel free economy.

Where to Study

Arizona State University
www.biodesign.asu.edu
Biodesign Institute
1001 S. McAllister Ave.
Tempe, AZ 85287-5001

California Institute of Technology
www.aphms.caltech.edu
Applied Physics & Material Science
1200 East California Blvd.
Pasadena, CA 91125

Colorado School of Mines
www.coloradofuelcellcenter.org
Colorado Fuel Cell Center
1310 Maple Street
Golden, CO 80401

Columbia University
www.seas.columbia.edu/earth
Earth Engineering Center
New York, NY 10027

Cornell University
www.ccmr.cornell.edu/industry/fuelcells
Cornell Fuel Cell Institute
330 Bard Hall
Ithaca, NY 14853

Fresno City College
www.fresnocitycollege.edu
Advanced Transportation Technology
1101 E. University Avenue
Fresno, CA 93741

National Fuel Cell Research Center
www.nfcrc.uci.edu
University of California, Irvine
Irvine, CA 92697-3550

Pittsburg State University
www.pittstate.edu/department/engineering-tech
Dept. of Engineering Technology
Pittsburg, KS 66762

Princeton University
www.princeton.edu/~energy/
Energy Group
Guyot Hall, Washington Rd.
Princeton, NJ 08544-1003

Tennessee Technological University
www.tntech.edu/cmr/
Center for Manufacturing Research
P.O. Box 5077
Cookeville, TN 38505

University of California, Davis
http://mae.engr.ucdavis.edu/~hypaul/
Hydrogen Production and Utilization
Laboratory
1 Shields Avenue
Davis, CA 95616

University of Minnesota
environment.umn.edu/iree/
Initiative for Renewable Energy and
the Environment
1954 Buford Avenue
St. Paul, MN 55108

Virginia Polytechnic Institute and State University
www.vtti.vt.edu/research/caar/
Center for Automotive Fuel Cell Systems
100S Randolph Hall
Blacksburg, VA 24060

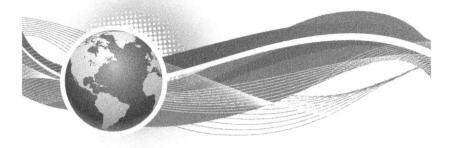

Green Transportation

In 1980, the minimum wage in the United States was $3.10. In 2013, the average was $7.25. The buying power of $3.10 in 1980 was, however, worth $7.82 in today's dollars, making today's minimum wage seem—well, not much of a deal.

In 1980, according to the Bureau of Transportation Statistics, the average passenger car got 24.3 miles per gallon. In 2006, that figure had climbed to 30 mpg—though the sales of passenger cars have declined compared to sports utility vehicles and light trucks, which means that the overall average is about the same.

NREL Electric Vehicle Grid-Integration Team reads data from test vehicles at the Vehicle Testing and Integration facility (VTIF). Photo: Dennis Schroeder, NREL

It's comparing apples to oranges, of course, but that 30 mpg figure is one we can all be proud of—while at the same time recognizing that there's still much more we can do to improve the fuel efficiency of our vehicles, just as we can do more to improve the way workers are paid in this country.

The advances of the last quarter-century have three big driving forces, if you'll forgive the pun. The first is the work of smart designers, engineers, and automakers, who figured out ways to make cars leaner and greener without compromising on safety or drivability. (Most of today's cars are in every way improvements on those of 1980 as a result.) The second is the power of government at all levels, municipal to national, to force those efficiencies. The third is the even greater power of the wallet, by which consumers have rewarded companies that have made better vehicles by buying them, even though the cost of a new car in 2012, adding finance charges and taxes, averaged a hair more than $30,000 for a passenger vehicle or light truck, while, according to the U.S. Department of Energy, in 1980 the figure was $7,600—that is, $19,176 in today's dollars.

The same forces are at work in making green vehicles today, but the opportunities are ever greater. Consumers, automakers, and governments alike are clamoring for better, safer, more fuel-efficient (and, of course, cheaper) vehicles. There's no better time to be involved in the many fields that contribute to that end, and no better time to be conscious of the environmental cost of transportation.

Careers in Green Transportation

Consider how important transportation is in our daily lives. Rightly or wrongly, we rely on long-distance transport to bring us much of what we eat and wear. We live in towns and cities where, for almost all of us, at least a few miles separate our homes from our workplaces, schools, shops, and other important points in our personal geographies. We've grown up in an era in which the automobile takes a place of pride in the very center of existence: it's not just a tool, but a means of personal liberation, self-fulfillment, escape, and fun.

Now, think about a green future in which cars run clean—or at least cleaner. Those of us who still use cars will find this a blessing, of course. But we'll also have learned to walk again, having remade our cities so that we can use our feet to get from one place to another, one reason for which fewer and fewer Americans are bothering to learn to drive these days. We'll take public transportation. We'll ride our bikes.

And some of us, of course, will take our hydrogen-powered vehicles down to the race track...

In that future, as one energy analyst notes, there'll be plenty of transportation-related jobs:

- Car and truck mechanic jobs
- Production jobs
- Bicycle repair and bike delivery services
- Gas-station jobs related to biodiesel
- Light rail system designers
- Public transit jobs related to driving, maintenance, and repair
- Energy retrofits to increase energy efficiency and conservation
- Alternative energy equipment installation
- Manufacturing jobs related to large-scale production of appropriate technologies
- Materials reuse

The list goes on and on. For the near future, most work in automotive engineering and design takes place within the automotive industry itself, with jobs typically requiring college or technical-school training. For a list of some of those schools, see the following section.

Researchers are needed in many different fields related to transportation. Within the Department of Energy, for example, the Advanced Vehicle Systems Research Program (http://peepsrc.ornl.gov/program_avs.shtml) conducts R&D on ancillary loads reduction, vehicle systems analysis, energy storage, power and propulsion systems, and advanced power electronics. Other groups of experts evaluate the performance of new vehicle and fuel technologies as they move into commercial use and work with industry to develop technical solutions and support to overcome problems that may result. Still others apply their expertise in fuels, lubricants, and emission control system technologies to guide relevant R&D in support of activities such as the Advanced Petroleum Based Fuels project.

Add to these technical and research fields the need for road and city planners, policy specialists, environmental specialists to conduct environmental-impact studies and to monitor environmental compliance, lawyers, managers, accountants, technical writers and editors, public-relations specialists, advocates, and salespeople—and that list goes on, too—and you'll see that there is much room in green transportation for a wide variety of skills, talents, and wishes.

Knowledge Is Power

Automotive engineering is a highly specialized enterprise, and it is no surprise that one of the flagship schools for it is the University of Michigan, whose Ann Arbor campus lies in the heartland of the automotive empire Henry Ford founded a century ago. "We are tasked with the most important objective of the transportation community: to educate the next generation automotive engineers who will achieve the changes required to survive in this competitive, global industry," writes Automotive Engineering Program director Margaret Wooldridge. "We strive to provide the best education on the most current topics important to the future of automotive engineering. Our goal is to create the individuals who will be the instruments of change." For an overview of UM's rigorous program, see www.interpro-academics.engin.umich.edu/auto/courses.htm.

Many universities and community colleges offer programs in automotive engineering and automotive engineering technology. *U.S. News and World Report* (*www.usnews.com*) has given these schools high marks in recent years:

Andrews University *www.andrews.edu*

Benjamin Franklin Institute of Technology *www.bfit.edu*

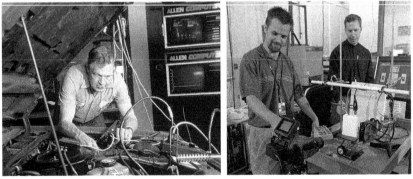

Left: A technician performs an inspection of an aftermarket conversion installation. Right: NREL engineers test a lithium-ion battery cell to improve performance of electric and hybrid vehicles. Photos: Natural Fuels Corporation; Warren Gretz, NREL.

Central Michigan University *www.cmich.edu*

Colorado State University, Pueblo *www.colostate-pueblo.edu*

Ferris State University *www.ferris.edu*

Indiana State University *www.indstate.edu*

Minnesota State University—Mankato *www.mnsu.edu*

Pennsylvania College of Technology *www.pct.edu*

Pittsburg State University *www.pittstate.edu*

SUNY—Farmingdale State College *www.farmingdale.edu*

Southern Illinois University *www.siu.edu*

Weber State University *www.weber.edu*

Western Michigan University *www.wmich.edu*

As well, the University of Cincinnati (*www.uc.edu*) offers a concentration in automotive design through its mechanical engineering program. In 2002, students there redesigned GM's EV1 (electrical vehicle 1), its first-generation electric car. ❖

Before Tesla, one startup idea Martin and Marc considered was smart sprinkler heads to aid in water conservation. They decided to conserve oil instead. Their Tesla Roadster is a high-performance electric car that can travel about 245 miles on one charge; 0 to 60 mph in less than 4 seconds with a top speed of 125 mph. It uses no gasoline and has zero emissions. Photo: Tesla Motors (www.teslamotors.com)

Resources

Advanced Vehicles & Fuels Research (NREL)

www.nrel.gov/vehiclesandfuels/
Working in partnership with public and private organizations, NREL researches, develops and demonstrates innovative vehicle and fuel technologies that reduce the nation's dependence on imported oil and improve our energy security and air quality.

Electric Drive Transportation Association

www.electricdrive.org
EDTA is dedicated to advancing electric drive as a core technology on the road to sustainable mobility. Their membership includes vehicle and equipment manufacturers, energy providers, component suppliers and end users.

Natural Resources Defense Council

www.nrdc.org/energy
The NRDC, a prominent environmental lobbying group, has a very useful web site with in-depth articles on energy and transportation issues.

Rocky Mountain Institute

www.rmi.org
Founded by visionary Amory Lovins, the Rocky Mountain Institute is one of the nation's leading think tanks devoted to energy efficiency and renewable energy sources. The RMI web site offers a broad-ranging set of documents on transportation, available for download.

U.S. Department of Energy

http://energy.gov/eere/transportation/vehicles
Numerous DOE programs focus on developing more energy efficient and environmentally friendly highway transportation technologies. A portal site follows, leading to several other sites within DOE and elsewhere.

Automakers such as General Motors (*www.gm.com*) and Toyota (*www.toyota.com*) also offer a wealth of information on their websites. Search online for your favorite manufacturer.

Honda's Fuel Cell Scooter equipped with Honda FC Stack. Photo: Honda

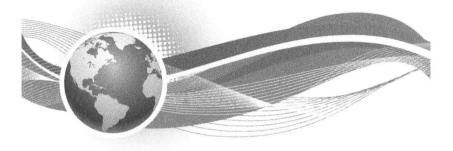

Energy Education & Economics

Teaching Energy

Renewable energy is a fascinating subject. Better put, it is a fascinating set of subjects, calling on knowledge that ranges from the sciences to communications to economics (and home economics) and beyond, on skills as various as climbing a ladder and driving a nail to aligning a wind-turbine propeller and tracking the sun's path.

Bringing this set of subjects to students is at once challenging and easy: Challenging because curricula are only now being developed in many aspects of renewable energy, easy because students at every level find the art, science, and industry of renewable energy to be accessible, interesting, and applicable to their daily lives.

NREL's Junior Solar Sprint and Lithium Ion Battery car competitions hosted by the U.S. Department of Energy's National Renewable Energy Laboratory. Photo: Dennis Schroeder, NREL

Ideally, a teacher bringing lessons and courses to students will share that view. His or her work is getting a little easier, too, for school districts are increasingly going green, and textbooks and other materials are daily arriving on the scene.

Teachers, regrettably, are a little slower to arrive: there just aren't enough of them, particularly in math and the sciences, and the need for teachers is expected to become greater and greater as retirement and attrition open up positions nationwide. Indeed, by 2010, more than 2.2 million teaching positions went unfilled, from elementary school to university. In part that was a result of the economic downturn and the regrettable—and appalling—tack many policymakers have been taking in defunding public education, a shameful refusal to invest in the nation's future. In part, though, it was also simply a lack of trained educators in so many fields and at so many levels, since, for many reasons, teaching is not for everyone.

It takes talent and patience to be a teacher, of course; it also takes knowledge, practice, and skill. The pay is often lower than in other sectors, and the stresses of keeping up with many dozens of students a day can take their toll. So, too, can the rigors of keeping up with new knowledge, changing technologies, and evolving curricula. But, as every teacher can tell you, connecting to that one promising student and setting him or her off on a world-changing course can readily outweigh the negatives—and if we are to change the way we make and consume energy and other resources, it will be thanks to bright students pointed in the right direction by excellent teachers. With luck, those teachers will be supported by their home institutions and allowed to innovate in order to teach about new technologies to the highest standards. That sort of educational innovation lies at the heart of the

Five Things Every High-School Graduate Should Know About Energy

What is electricity and where does it come from?

What are the sources of renewable and non-renewable energy?

What are the environmental and economic impacts of our energy use?

What does it mean to move from a carbon based to a hydrogen economy?

What will the energy portfolio in the future look like?

SOURCE: NREL OFFICE OF EDUCATION

most interesting schools in the country, among them San Diego's High Tech High *(www.hightechhigh.org)* and MIT's Media Lab *(www.media.mit. edu)*, but it can be imported everywhere that administrators allow it to flourish—no small matter in a time when so many teachers are forced to teach to the test and not to the true subject at hand.

What follows are some resources, many of which can be adapted to early learners as well as collegians. Whether you're a teacher or a student, you'll find plenty to engage your interest here.

If the future seems uncertain for schools as we now know them, one thing that seems unshakably true is that what is always required for innovation to happen is education—and without the possibility of education, things can never improve, progress can never be made. Remarked Rensselaer Polytechnic Institute President Shirley Ann Jackson in a recent interview, "I've never seen any innovation yet that just popped out of the air. It comes from people's ideas. So if we don't have the right talent, we're not going to be able to meet our energy needs."

Education, to put it another way, is the single most important fuel source to power a renewable future.

Requirements for Teaching Energy

Middle school teachers (grades 4–8) and high school teachers generally specialize in a specific subject, with renewable energy falling under the various sciences, such as earth and space science, life science, physical science, as well as technology / engineering. History, economics and social sciences are also applicable subjects for renewable energy and sustainable living.

Traditionally, public school teachers were required to have at least a bachelor's degree, complete an approved teacher education program, and be licensed. Many states, however, are now offering alternative licensing programs to attract people who have bachelor's degrees in a particular subject, but lack the education courses required for a regular license. Private school teachers often have slightly different requirements.

Educators at colleges and universities usually have master's and doctorate degrees, while vocational/technical instructors typically need work-related experience, and often a license or certificate, in their field.

Department of Energy Supports Education

NREL's Education Programs

"By 2050, the world will be struggling to find an additional 20 terawatts of energy and the U.S. will be competing along with countries such as China and India to satisfy our growing energy needs. In order to meet the challenges of this new energy future, we will need new ways of thinking about and producing energy. Here at NREL, we believe that performing renewable energy and energy efficiency R&D is only part of the equation. Education is key to creating a new energy future.

"From elementary school science mentoring to senior-level research participant programs, NREL's educational opportunities help provide the link to this new future. Our K-12 science programs engage young minds in renewable energy and also provide support for teachers. Our college and post-graduate programs help to develop a capable and diverse workforce for the future through mentored research internships and fellowships. And our more senior-level programs—from post-doctoral researchers up to sabbatical and faculty appointments—provide the opportunity to participate in the Laboratory's research and development programs, initiate new areas of research, and establish a base for ongoing collaborations between NREL and our stakeholders."

www.nrel.gov/education

Academies Creating Teacher Scientists

The Department of Energy Academies Creating Teacher Scientists (DOE ACTS) program is designed by the Office of Science to create a cadre of outstanding science and math teachers with the proper content knowledge and scientific research experience to serve as leaders and agents of positive change in their local and regional teaching communities. This three-year program uses the unmatched wealth of mentoring talent at the DOE National Laboratories to guide and enrich the teachers' understanding of the scientific and technological world.

www.dep.anl.gov/p_k-12/acts

Education Resources

Alliance to Save Energy

www.ase.org

This organization offers educators a wide range of tools and resources to bring energy efficiency into the classroom while helping students build vital real-world skills. Its web site offers hundreds of lesson plans.

Discovery Channel Curriculum Center

school.discoveryeducation.com/curriculumcenter

A trove of information for students and teachers alike, their web site has a curriculum center with lesson plans in all sorts of energy-related fields.

EducationNation.com

www.educationnation.com

Funded in part by General Motors and maintained by NBC, EducationNation.com is a clearinghouse with numerous resources for teachers and learners in a variety of fields, including energy.

Energy Education & Workforce Development, U.S. Department of Energy

www1.eere.energy.gov/education/

This is a superb source of materials for teaching about energy. Students who come to the site can also find ideas for energy-related science projects and resources to help them write reports on energy topics. High-school and college students can also find information on energy careers, energy-related college degrees and programs, scholarships, energy-related internships, and job listings.

Energy for Educators

www.energyforeducators.org

A website hosted by the Idaho National Laboratory and geared to Idaho curriculum standards, this site has material of much use to teachers everywhere, including a full set of K-12 STEM lesson plans. All are of good quality, though the site administrators warn that they have not been peer-reviewed.

Energy Information Administration

www.eia.doe.gov

EIA offers a wealth of information on all aspects of renewable energy, including reports and resources for professionals and students alike. The web site contains links to energy-education materials for students and teachers.

Energy Quest

www.energyquest.ca.gov

This award-winning energy-education web site of the California Energy Commission offers abundant resources for budding scientists, inventors, and energy consumers, who will find the web site full of information on energy production and conservation. Of particular use is the section devoted to student-teacher resources.

Energy Teaching Resources

www.environment.nsw.gov.au/sustainableschools/teach/energyteach.htm

The government of New South Wales, Australia, offers a vigorous program of teaching materials for renewable energy at this must-visit website.

Florida Solar Energy Center

www.fsec.ucf.edu
FSEC works with educators to engage and interest students of all levels to help create a brighter energy future. They offer several K-12 curriculum units online, as well as continuing education coursework for teachers.
1679 Clearlake Road
Cocoa, FL 32922-5703
(321) 638-1000

Foundation for Water and Energy Education

http://fwee.org
Devoted to hydropower, this site has numerous resources for teachers, including a hands-on science curriculum program for grades 6–8.

Green Teacher Magazine

www.greenteacher.com
P.O. Box 452
Niagara Falls, NY 14304-0452
(416) 960-1244

National Energy Education Development Project (NEED)

www.need.org
NEED offers newsletters, the monthly publication Career Currents, classroom kits, and many other resources. Its web site is a mine of good information.

National Energy Foundation

www.nef1.org
NEF is a non-profit organization devoted to innovative teacher training and student programs.

National Renewable Energy Laboratory Education Program

www.nrel.gov/education
The NREL Education Program web site is a must-bookmark destination for teachers and students with an interest in all things having to do with energy. The site lists scholarships, grants, internships, and jobs, and it provides abundant resources for lesson planning and curriculum building.

National Science Teachers Association

www.nsta.org
NSTA, with 55,000 members, is a clearinghouse that helps science teachers connect with one another. They provide valuable resources, including lesson plans and reference books.

Solar in the Schools

www.solarenergy.org/solar-schools
Solar Energy International's SIS program targets grade school youth and focuses on experiential learning of energy concepts. SIS emphasizes critical thinking skills by exploring energy choices, costs and solutions within communities.
520 S. Third Street, Room 16
Carbondale, CO 81623
(970) 963-8855

Teachers First

www.teachersfirst.com
This well-organized site contains lesson plans and other materials on nearly every topic that touches on renewable energy.

Teaching About Energy in Geosciences Courses

serc.carleton.edu/NAGTWorkshops/energy/
Carleton University offers a wealth of resources, many quite advanced, on energy and pedagogy, including developing standards for energy literacy.

Union of Concerned Scientists, Renewables Are Ready

www.ucsusa.org/assets/documents/clean_energy/renewablesready_fullreport.pdf

A research report and curriculum guide for energy education in middle and high school, this UCS publication was last updated in 2003 and could use a little freshening, but still has value.

University of Wisconsin K-12 Energy Education Program (KEEP)

www4.uwsp.edu/cnr/wcee/keep

KEEP's mission is to initiate and facilitate the development, dissemination, implementation, and evaluation of energy education programs within Wisconsin schools. Teachers outside Wisconsin will find much of use on its website, however.

U.S. Department of Energy: Energy Literacy

www1.eere.energy.gov/education/pdfs/energy_literacy_1_1_high_res.pdf

Though the last listing here, this is the place to start.

· ·

Taking Science and Renewable Energy on the Road

The Leo on Wheels (LOW) is an educational outreach program from The Leonardo science, technology, and art museum in Salt Lake City. LOW travels to middle schools throughout Utah, serving nearly 14,000 students each year. It integrates science core content with 21st century thinking skills through hands-on, inquiry-based activities that foster critical thinking, creativity, and innovation. *www.theleonardo.org/LOW*

· ·

Energy Economics

Economics, the old saw has it, is the dismal science: It is a matter of numbers, projections, pie charts, tables. A new generation of economists has helped reverse some of that dismal quality, though, by wedding the theories and concepts that economists use to think with—opportunity cost, externality, marginal utility—to the things that actual humans actually do. It's probably no surprise that those things are less rational than we might like to think, but there economists can be helpful, too, by pointing out rational courses of action in times of scarcity—and charting a course of renewable energy is nothing if not rational.

Economists who specialize in energy are becoming ever more influential in shaping that course. Now, the basics of economics are going to be the same no matter where they are studied, with a few variations (the economics learned at Chicago is likely to be somewhat more conservative, for example, than the economics learned at Berkeley). That basic economic course is going to consist of studies in microeconomics, macroeconomics, and various branches of theory.

Beyond that lies graduate school, which is essential for most work in energy economics. And there lie more significant differences. At Berkeley,

to name just one school, an emphasis has developed that takes into account not just economics, but also what energy economist Derek Lemoine deems "what you really need to know"—that is to say, how electricity works, how utility grids are constructed and expanded upon, the differences between renewable and nonrenewable energy, and the fiscal policies that guide each. In some schools, coursework is geared to the regulatory climate, in others to energy markets (with those latter often pitched to students already possessing an MBA). At Stanford, programs have developed in the economics of energy efficiency; at Carnegie Mellon, in the hybrid energy economy.

"If I had to pick one program that would allow me to be hired as an energy economist," says Lemoine, an assistant professor of economics at the University of Arizona, "it would be Berkeley." A degree from Berkeley, as one might expect, is a license to teach others in energy economics. In that academic spirit, he recommends Cal Tech, Stanford, Boston University, Carnegie Mellon, and the University of Maryland (the last for its Joint Global Change Institute).

Academic economists do more than teach others economics, of course. One way in which researchers help is to frame questions that transcend dollars and cents and move into more philosophical realms. Economists assume, for instance, that economies will always grow over the long run, and that the generations that succeed our own will be richer than we are— at least that's the way it has worked, averaging out the occasional bumps, for centuries. If that is so, then consider this: Future generations will suffer more from climate change than we are suffering. But if they are richer than we are, should we act now to remove or reduce the causes of future harm, or should we let them pay for the solutions to our problems? Economists help us think through such conundrums.

On a more immediately practical level, many other programs are available that lead to jobs in the private sector and the public sector outside academia. For instance, Cornell University offers an innovative program in energy economics and engineering whose graduates go on to careers in chemical engineering, management, and policy. The one-year program leads to a master's degree in engineering (M.Eng), and its coursework is as follows:

Fall Semester: 15 credits, including CHEME 6660 Analysis of Sustainable Systems+Modules, CHEME 6640 Energy Economics, Economic and Social Impact Elective, and Business Practices Option.

Spring Semester: 15 credits, including CHEME 6650 Energy Engineering, Environmental Impact Elective, Energy Technology Elective, CHEME 5650 Energy-related M. Eng. Project.

The Colorado School of Mines offers M.S. and Ph.D. programs in energy economics and minerals that, while by definition not directly involving renewables, teach concepts that can be applied to renewable energy—and, in any case, minerals are required to produce the renewable infrastructure. As the school describes it, there are two basic tracks in its course offerings: an operations research/operations methods track to "provide the student with increased competence in the key functional areas of business to meet the challenges of optimization and value creation in the competitive global mineral and energy industries," and an applied economics track, to "teach students to analyze markets and public policies affecting mineral, energy and environmental issues." The latter field, the school adds, teaches proficiency in "applying the primary tools of economic analysis such as benefit-cost, investment and market analysis of consumers and producers."

Similarly, students in the University of Michigan's Erb Institute MBA/MS program earn both a master of business administration (MBA) from the Ross School of Business and a master of science (MS) from the School of Natural Resources and Environment. The program equips leaders, executives, and managers with the skills and knowledge necessary to create environmentally and economically sustainable organizations.

Says energy economist and venture capitalist Alan Salzman, "The world has spoken: It wants meaningful portions of its energy to be renewable." Getting the world to that point will require economists as much as technicians. To get an overview of the field of energy economics, have a look at the free MIT OpenCourseware offering in the field (http://ocw.mit.edu/courses/economics/14-44-energy-economics-spring-2007/index.htm). The course, as you'll see, was last taught in 2007, but the basics remain the same. ❖

Energy Economics Resources

Information Clearinghouses

Center for Energy Economics
University of Texas at Austin
www.beg.utexas.edu/energyecon

Joint Global Change Research Institute
www.globalchange.umd.edu

United States Association for Energy Economics
www.usaee.org

For Further Reading

Energy Economics: Concepts, Issues, Markets and Governance by Subhes C. Bhattacharyya (Springer, 2011).

Energy and the Wealth of Nations: Understanding the Biophysical Economy by Charles A.S. Hall (Springer, 2012).

Energy Finance: Analysis and Valuation, Risk Management, and the Future of Energy by Betty Simkins and Russell Simkins, eds. (Wiley, 2013).

Where to Study

California Institute of Technology
www.caltech.edu
1200 East California Blvd.
Pasadena, CA 91125
(626) 395-6811

Carnegie Mellon University
www.cmu.edu
5000 Forbes Ave.
Pittsburgh, PA 15213
(412) 268-2000

Colorado School of Mines
www.mines.edu
1500 Illinois St.
Golden, CO 80401
(303) 273-3000

Cornell University
www.cheme.cornell.edu/academics/ graduate/meng/specializations/eee.cfm
School of Chemical and Biomolecular Engineering
Ithaca, NY 14850
(607) 255-5241

Massachusetts Institute of Technology (MIT)
web.mit.edu
77 Massachusetts Avenue
Cambridge, MA 02139-4307
(617) 253-1000

Stanford University
www.stanford.edu
450 Serra Mall
Stanford, CA 94305–2004
(650) 723-2300

University of California Berkeley
www.berkeley.edu
101 Sproul Hall
Berkeley, CA 94704
(510) 642-6000

University of Michigan
www.erb.umich.edu
Erb Institute for Global Sustainable Enterprise
440 Church Street
Ann Arbor, MI 48109–1041
(734) 647-9799

Appendix

Thank you

...to all the companies and organizations who so generously supplied photos and graphics for this book, especially the National Renewable Energy Lab. To Scott Sklar, David Petroy, and Michael Sykes for sharing their hard-earned insights. And to Linda Lung, Joe Jordan, and Rex Ewing for their technical advice regarding the many facets of renewable energy.

Where to Study

According to the Digest of Education Statistics, published annually by the U.S. Department of Education, as of the end of the 2012 there were 4,599 degree-granting public and private community colleges, colleges, and universities in the United States (along with 89 others in U.S. territories, almost all of them in Puerto Rico). California led the nation in the number of these postsecondary-education institutions with 454 of them, followed by New York, Pennsylvania, Texas, and Ohio. The lowest-ranked of them was Alaska, which is no surprise at all, given its small population.

The point is that no state is without numerous choices for study after high school. Add to those 4,599 institutions the nation's more than 2,300 vocational schools, and that gives students a wealth of choices that no other country enjoys. That is especially true of the cluster of fields that this book addresses, all falling under the rubric of renewable energy.

What to make of all these choices? Many factors will go into where you decide to study. Do you want to work as a technician? If so, a two-year college, trade school, or apprenticeship will probably fill your needs very well. Do you want to conduct scientific research in, say, hydrogen fuel cell technology? If so, you'll almost certainly need a doctorate in engineering, materials science, or a related field. Would you like to work in policy, helping move the cause of renewable energy along in the public and private spheres? Then some mix of study in planning, policy, administration, law, and science is in order. Would you like to be an evange-

list for wind power? You may find that training in sales, business, and communications comes in handy, as well as a solid grounding in how wind turbines work.

Such questions are important. So are others, such as how much time you'd like to spend in school, how much education you can afford (particularly given the long-lasting burden of student loans), where you'd like to study and live, and where the best job opportunities are. The answers to them will have much bearing on your course.

In the preceding chapters, we've provided pointers to some of the best web sites and other resources in individual fields such as geothermal energy and green building. Here we'll list several dozen institutions that have distinguished themselves in some aspect or another of renewable energy. For a more directed search, have a look at the U.S. Department of Education's College Navigator (*nces.ed.gov/collegenavigator*), a terrific resource, and at some of the college guidebooks and trade-school listings online. Don't be afraid to ask for help, either—we can almost guarantee you that your local librarian will be delighted to assist you in your quest for information.

Arizona State University

schoolofsustainability.asu.edu
School of Sustainability
PO Box 87511
Tempe, AZ 85287-2511
(480) 727-6963

The recently established School of Sustainability at Arizona State University offers courses in sustainable environmental resources, resource allocation, energy and material use, engineering, and many other subjects. ASU's College of Technical and Applied Science also offers courses in renewable and alternate energy sources, including fuel cells (see technology.poly.asu.edu).

Boise State University

http://earth.boisestate.edu
Department of Geosciences
1910 University Drive
Boise, Idaho 83725-1535
(208) 426-1631

The Department of Geosciences of Boise State University, located in Idaho's capital, describes its program as follows: "Our research seeks not only to advance understanding of the surface, near surface, and deep Earth environments, but also to produce science that addresses societally relevant problems such as climate change, human-environment interactions, alternative energy sources, and basic materials." Boise State is also the center of the federally funded National Geothermal Data System (NGDS), offering opportunities for internships and graduate-level research.

California Institute of Technology

www.caltech.edu
1200 East California Blvd.
Pasadena, CA 91125
(626) 395-6811

Caltech is a leading research institution in the study of renewable energy, especially solar energy and biofuels. Its chemistry department, to name just one program, sponsors research in solar energy conversion and storage, methane oxidation, and nitrogen fixation.

California Polytechnic State University

www.calpoly.edu
College of Agriculture, Food and Environmental Sciences
San Luis Obispo, CA 93407
(805) 756-2161

The BioResource and Agricultural Engineering major at Cal Poly offers hands-on experience in a wide range of engineering skill areas, renewable energy and waste treatment, electronics and control systems, and resource information systems. The BRAE program is accredited by the Accreditation Board for Engineering and Technology (ABET).

Coconino Community College

www.coconino.edu
Lone Tree Campus
2800 S. Lone Tree Road
Flagstaff, AZ 86001-2701
(928) 527-1222, (800) 350-7122

CCC offers a certification program as an alternative energy technician, with an intermediate certificate requiring 28–29 credit hours and an advanced certificate requiring 52–57 credit hours.

Colorado State University

www.warnercnr.colostate.edu
Warner College of Natural Resources
103 Natural Resources Building
Fort Collins, CO 80523
(970) 491-5629

Warner College of Natural Resources is committed to offering a comprehensive range of undergraduate and graduate degree programs that directly address today's most important environmental and natural resource issues. Its programs are grounded in state-of-the-art science and technologies and involve students in direct problem-solving experiences. Its students are well prepared to become leaders in environmental and natural resources management and science.

Columbia University

www.me.columbia.edu
2960 Broadway
New York, NY 10027-6902
(212) 854-1754

A leading university with undergraduate and graduate instruction in every field that touches on renewable energy, Columbia is home to a highly ranked Department of Mechanical Engineering that offers courses in energy sources, materials science, electrochemistry, and energy systems. See also *www.cheme.columbia.edu*

Cornell University

www.geo.cornell.edu/eas/energy
College of Engineering
Ithaca, NY 14853
(607) 255-5241

Cornell University is a leading center for the study of renewable energy. Its College of Engineering offers minors for undergraduate and graduate students in Sustainable Energy Systems, including geothermal energy.

Drexel University

www.chemeng.drexel.edu
Dept. of Chemical and Biological Engineering
3141 Chestnut Street
Philadelphia, PA 19104
(215) 895-2227

Development of technologies for harnessing energy from renewable resources and for reducing the environmental impact of worldwide resource consumption is now receiving major national and international attention. Chemical and Biological Engineering Department faculty at Drexel University are leading innovative research programs related to energy and the environment, including the production of biodiesel from natural and waste oils containing high free-fatty-acid contents, developing new materials for higher efficiency solar cells, deriving polymers from renewable fatty acid monomer, and modifying the composition and structure of fuel cell catalyst layers for better performance.

Fitchburg State University

www.fitchburgstate.edu
160 Pearl Street
Fitchburg MA 01420–2697
(978) 665-3000

Fitchburg State offers a range of courses in energy-related fields, including programs in industrial technology, architectural technology, energy management technology, and facilities management. The architectural technology concentration offers the study of technical systems in architecture, including energy systems, and ends with the development of professional practices.

Florida A&M University

www.famu.edu

College of Agriculture & Food Sciences
Tallahassee, FL 32307
(850) 599-3000

Florida has the most abundant biomass resources in the United States. Florida A&M, a historically black college (now university), offers a range of programs in bioenergy education, holding that it can be the catalyst for a new suite of industries in the state.

Georgia Institute of Technology

www.gatech.edu

North Avenue
Atlanta, GA 30332
(404) 894-2000

Georgia Institute of Technology is a national leader in energy research, offering undergraduate and graduate courses in industrial systems engineering, civil engineering, and sustainability. It has recently paired with private companies in the region to study the feasibility of offshore wind installations on the Georgia coast.

Illinois Institute of Technology

www.iit.edu

3300 South Federal Street
Chicago, IL 60616-3793
(312) 567-3000

Energy/Environment/Economics

E3 is an academic program of research and coursework for students in chemical, mechanical, environmental, and electrical engineering. The research program encompasses areas of specialization that relate to energy, sustainable development, industrial ecology, and environmental design.

Wanger Institute for Sustainable Energy Research

www.iit.edu/wiser/

Students and researchers at Illinois Institute of Technology (IIT) are striving to improve the quality of life in our nation while preserving the natural resources and the environment for future generations. At the Wanger Institute for Sustainable Energy Research (WISER), more than 50 faculty members and their students are currently involved in energy and sustainability research and educational activities across the colleges and institutes at IIT.

Iowa Lakes Community College

www.iowalakes.edu

300 South 18th Street
Estherville, IA
(712) 362-2604 or (800) 521-5054

Iowa Lakes Community College's Wind Energy and Turbine Technology Program offers a two-year associate in applied science program that helps meet a growing demand for skilled technicians who can install, maintain, and service modern wind turbines. The diploma program consists of three terms of coursework, providing training in the construction, maintenance, and operation of wind turbines. A second year of coursework leads to the associate's degree, with additional training in the diagnosis of turbines, computerized control and monitoring systems, wind turbine siting, and data acquisition.

Iowa State University

www.windenergy.iastate.edu/reu.asp

Wind Energy Science, Engineering, and Policy (WESEP)
College of Engineering
104 Marston Hall
Ames, IA 50011
(515) 294-1588

Iowa State University offers an intensive ten-week on-campus research program in Wind Energy Science, Engineering, and Policy (WESEP) for undergraduate students. Ten fellowships are sponsored each year by the National Science Foundation's (NSF) Research Experiences for Undergraduates (REU) program. Students work collaboratively in interdisciplinary teams with faculty and graduate students to receive training and get hands-on research experience in areas that address critical, long-term national needs in wind energy-related areas.

Lane Community College

www.lanecc.edu/sustainability/energy-management-program
4000 East 30th Avenue
Eugene, OR 97405
(541) 463-3000

See the chapter on Energy Management for a description of LCC's extensive programs.

MIT Laboratory for Energy and the Environment

http://web.mit.edu/mitei/lfee/
Contact: Teresa Hill, Ph.D.
MIT Room E19-370U
Cambridge, MA 02139
(617) 253-1341

The Laboratory for Energy and the Environment (LFEE), an integral part of the MIT Energy Initiative (MITEI), fosters collaboration among industry, government, academia, nongovernmental organizations, and the public to address not only the complex interrelationships between energy and the environment, but also the technological, economic, and social aspects of sustainable energy development and use. Graduate fellowships at MIT are vital support for MIT's interdisciplinary research in energy, the environment, and sustainability topics.

Michigan State University

canr.msu.edu
College of Agriculture & Natural Resources
102 Agriculture Hall
East Lansing, MI 48824
(517) 355-0232

Michigan State, an old and well-regarded land-grant college, is the home of the Biomass Conversion Research Laboratory. Its agriculture college offers several graduate, undergraduate, and two-year certificate programs in fields that pertain to renewable energy, including biosystems and agricultural engineering, construction management, crop and soil sciences, forestry, and agricultural technology.

Montana State University

www.sfbs.montana.edu
Sustainable Food & Bioenergy Systems
Bozeman, MT 59717
(406) 994-5640

Montana State University's SFBS program offers interdisciplinary undergraduate degrees in four concentrations: the Sustainable Food Systems Option, Sustainable Crop Production Option, Agroecology Option, and Sustainable Livestock Production Option. Each option includes coursework in sustainable food and bioenergy production.

Montana Tech of the University of Montana

www.mtech.edu
1300 West Park Street
Butte, MT 59701
(800) 445-8324

Offering two- and four-year degrees in both trade and academic fields, Montana Tech has a well-regarded program in geological and geophysical engineering and is located in one of America's most active areas for geothermal energy.

North Carolina State University

ncsc.ncsu.edu
North Carolina Solar Center
1575 Varsity Drive
North Carolina State University
Raleigh, NC 27606
(919) 515-3480

Created in 1988, the North Carolina Solar Center serves as a clearinghouse for solar and other renewable energy programs, information, research, technical assistance, and training for the citizens of North Carolina and beyond. Students at NCSU have opportunities to learn at the center, and the school also offers excellent programs in many other renewable energy technologies.

Norwegian University of Science and Technology (NTNU)

www.ntnu.edu/studies/msb1
Department of Hydraulic and Environmental Engineering
attn. Ånund Killingtveit
NO-7491 Trondheim
Norway
(47) 73 59 47 51

Please see the Hydropower chapter for a description of the NTNU international master's degree program.

Oklahoma City Community College

www.okc.cc.ok.us
7777 South May Avenue
Oklahoma City, OK 73159-4444
(405) 682-1611

OKCC's program prepares individuals to apply basic engineering principles and technical skills in support of engineers and other professionals engaged in developing solar-powered energy systems. Includes instruction in solar energy principles, energy storage and transfer technologies, testing and inspection procedures, system maintenance procedures, and report preparation.

Oklahoma State University

www.hvac.okstate.edu
Stillwater, OK 74078
(405) 744-5000

The Building and Environmental Thermal Systems Research Group of Oklahoma State University is made up of faculty members, students, and researchers with interests that include building heat transfer, HVAC systems modeling, building energy simulation, hydronic heating systems, geothermal heat pump systems, and ground loop heat exchanger technology.

Oregon Institute of Technology

www.oit.edu

With campuses in Klamath Falls, Wilsonville, and La Grande, as well as an out-of-state extension in cooperation with Boeing Aircraft in Seattle, OIT is home to the Geo-Heat Center, a research institution that "provides technical analysis for those actively involved in geothermal development." OIT also offers an undergraduate renewable energy degree, which includes a course in geothermal energy and ground-source heat pumps. Online courses also available.

Oregon State University

www.oregonstate.edu
Corvallis, OR 97331–4501
(541) 737-1000

Oregon State offers more than two hundred undergraduate and one hundred graduate degree programs through its twelve colleges, including the University Honors College, one of only a handful of degree-granting honors programs in the United States. Numerous programs offered through the College of Engineering pertain to wind energy at both the undergraduate and graduate levels, while other departments and colleges provide training in related disciplines.

Pennsylvania College of Technology

www.pct.edu
One College Avenue
Williamsport, PA 17701
(570) 326-3761

PCT offers two- and four-year courses in solar technology, electric-power-generation technology, electrical technology, and other areas relevant to renewable energy.

Pennsylvania State University

www.bioenergy.psu.edu
Penn State Biomass Energy Center
225 Agricultural Engineering Building
University Park, PA 16802
(814) 865-3722

Penn State's Biomass Energy Center acts as a clearinghouse of information and a coordinator of campus-wide activities related to bioenergy. Its associated faculty come from several departments, and exploring their listings will yield information on a variety of programs. In addition, the Center offers short courses for continuing education and professional development.

Rensselaer Polytechnic Institute

www.rpi.edu/dept/ees/
Department of Earth and Environmental Sciences
Jonsson-Rowland Science Center, 1W19
110 8th Street
Troy, NY 12180
(518) 276-6474

RPI offers undergraduate and graduate training in the geosciences and has become a leading center of geothermal-energy studies. The programs include the study of Earth's component materials, the development of its structures and surface features, the processes by which these change with time, and the origin, discovery, and protection of its resources—water, fuels, and minerals.

Riverland Community College, Minnesota

www.riverland.edu

With campuses in Albert Lea, Austin, and Owatonna, Riverland's wind turbine technician diploma program offers students a new career path within the construction electrician and industrial maintenance and mechanics programs. This option gives students a choice of training to be an entry-level wind turbine technician, someone who plays a key role in ensuring quality, safety and service involving the operation and maintenance of wind turbine units, performing mechanical and electrical troubleshooting as well as repair and preventative maintenance.

San Juan College

www.sjc.cc.nm.us
4601 College Blvd.
Farmington, NM 87402
(505) 327-5705

The Renewable Energy Program gives the student a solid foundation in the fundamental design/installation techniques required to work with renewable technologies. The concentration in Passive Solar Design and Analysis is offered as an A.A.S. degree and/or a one-year certificate. Both avenues include hands-on electrical training in a modern computer-based laboratory setting where the National Electrical Code (NEC) is emphasized throughout the curriculum.

St. Cloud State University

www.stcloudstate.edu
Atmospheric and Hydrologic Sciences
720 4th Avenue South
Saint Cloud, MN 56301
(320) 308-3144

The hydrology major, leading to a BS degree, focuses on the quantitative study of surface and ground water and provides the background for a variety of entry level job opportunities in industry and government and for those intending to pursue graduate work in the field. Students are strongly urged to consider a minor in a related field such as meteorology, geography/geographic information science, or environmental studies. Coursework in the major involves environmental and earth sciences, chemistry, geology, mathematics, and physics.

Syracuse University

www.syr.edu
Syracuse, NY 13244
(315) 443-1870

The Building Energy and Environmental Systems Laboratory (beesl.syr.edu) in the Department of Mechanical and Aerospace Engineering is a key research lab. Its mission, in part, is to advance the science and develop innovative technologies in the areas of indoor environmental quality and building energy efficiency. BEESL's labs are used for studies of combined air, heat, moisture and contaminant transport through building envelopes; interactions between indoor, outdoor environments and HVAC systems/components; sensitivity, accuracy and reliability of environmental sensors and control systems; and many other topics of interest to energy managers and students of renewable energy.

Texas Tech University
Wind Science and Engineering Research Center

www.depts.ttu.edu/weweb/
3301 4th St
Lubbock, TX 79415
(806) 742-2011

Texas Tech University's newly formed National Wind Institute (NWI) is based on a strong foundation of more than four decades of research and edu-

cation on the impact of wind on structures and human life. TTU has created the NWI to better support the interdisciplinary research and educational opportunities in wind science, engineering, and energy.

University of Alaska Southeast

uas.alaska.edu
11120 Glacier Hwy.
Juneau, AK 99801
(907) 796-6000

The UAS offers a program that teaches the essentials of building diagnostic assessment and building durability, performance, and energy efficiency. One-year certificate programs are offered in residential/light construction, building energy retrofit technology, and power technology, among other energy-related fields.

University of California, Davis

http://hydscigrad.ucdavis.edu
Hydrologic Sciences Graduate Group
Davis, CA 95616
(530) 752-6810

UC Davis houses one of the nation's most esteemed graduate programs in hydrology. The Hydrologic Sciences Graduate Group (HSGG) offers MS and PhD degrees emphasizing physical, chemical, and biological processes that affect the circulation of water and solutes on Earth. Students with training in hydrologic sciences, geology, geophysics, engineering, soil science, biology, chemistry, computer science, fluid mechanics, mathematics, and physics are strongly encouraged to apply for admission to the graduate program. Coursework includes studies in geographic information systems, geology, mechanics, irrigation, fluid dynamics, limnology, environmental law, and many other subjects.

University of California, Los Angeles

www.ucla.edu
Department of Architecture
Los Angeles, CA 90095
(310) 454-7348

The internationally recognized UCLA architecture program offers a specialization in energy design and climate-responsive design. Credit is given for AIA continuing education; see *www.aud.ucla.edu/energy-design-tools* for more information.

University of California, Merced

naturalsciences.ucmerced.edu
5200 North Lake Road
Merced, CA 95343
(209) 228-4400

Merced's undergraduate major in Earth Systems Science is designed to provide students with a quantitative understanding of the physical, chemical, and biological principles that control the processes, reactions, and evolution of Earth. Core courses cover the fundamentals of chemistry, biology, hydrology, ecology, and earth sciences. Graduates are well prepared for either graduate studies or jobs in the areas of environmental science and conservation, ecosystem and natural resource management and science, and many aspects of agricultural sciences.

University of California, Santa Barbara

www.es.ucsb.edu/academics/hydo-sci-major/bs
Environmental Studies Program
4312 Bren Hall
UC Santa Barbara CA 93106–4160
(805) 893-2968

Please see the text for a description of the UCSB undergraduate degree in hydrologic sciences.

University of Central Florida

www.fsec.ucf.edu
Florida Solar Energy Center
1679 Clearlake Road
Cocoa, FL 32922–5703
(321) 638-1000

FSEC is the largest and most active state-supported renewable energy research institute in the United States. It is an accredited laboratory for testing and certification of solar technologies. Established in 1975, it also conducts research in building science, solar energy, hydrogen and alternative fuels, fuel cells, and other advanced energy technologies. The school offers courses and workshops that take advantage of this renowned resource.

University of Colorado Environmental Center

ecenter.colorado.edu
University Memorial Center, Rm 355
Boulder, CO 80309-0207
(303) 492-8308

Colorado is one of the greenest campuses in the nation, literally and figuratively, and a leader in the use of renewable energy sources such as wind and solar power. The university offers a wide range of programs in every aspect of renewable energy.

University of Florida

www.floridaenergy.ufl.edu
Florida Energy Extension Service
P.O. Box 118300
Gainesville, FL 32611-8300
(352) 392-0947

The University of Florida provides continuing education courses in green building for Florida licensed building contractors and architects, as well as other interested parties. The extension service involves other state universities and a network of institutions elsewhere in the South—and even a partnership with Yale University.

University of Illinois at Chicago

www.erc.uic.edu/cap/engsolutions.htm
Energy Resources Center
851 S. Morgan Street
Chicago, IL 60607–7054
(312) 996-4490

The Energy Resources Center is an interdisciplinary program whose areas include energy management assessments, economic modeling, analysis of policy and regulatory initiatives, and education. ERC provides energy auditor/energy efficiency training to a selected team of high-school students each year, as well as an energy camp that offers a week of more intense training to the same students.

University of Massachusetts
Center for Energy Efficiency and Renewable Energy

www.ceere.org
160 Governors Drive
Amherst, MA 01003
(413) 545-0684

The Center for Energy Efficiency and Renewable Energy (CEERE) provides technological and economic solutions to environmental problems resulting from energy production, industrial, manufacturing, and commercial activities, and land use practices. The research program is built upon four subgroups with unique abilities to service energy and environmental problems. CEERE offers research, training and educational experiences for graduate and undergraduate engineers and scientists.

University of Nevada, Reno

www.unr.edu
1664 N. Virginia Street
Reno, NV 89557–0042
(775) 784-1110

A land-grant institution with graduate programs in a wide range of fields, Nevada has a well-established program of study in renewable energy. The program includes students from Truckee Meadows Community College, using the vast, nearby Steamboat Geothermal Complex and Nevada's abundant sunshine as laboratories. For information on the specialized program of study under the rubric of the National Geothermal Academy, see www.unr.edu/geothermal/NGA.htm.

University of New South Wales

www.pv.unsw.edu.au
School of Photovoltaic and Renewable Energy Engineering
Electrical Engineering Building
Sydney NSW 2052
Australia
+61 2 9385 1000

The School of Photovoltaic and Renewable Energy Engineering delivers the world's first photovoltaic and renewable energy engineering degree programs. The School includes the Australian Research Council Centre for Advanced Silicon Photovoltaics and Photonics (also known as the ARC Photovoltaics Centre of Excellence).

University of Strathclyde

www.strath.ac.uk/windenergy
UK Wind Energy Research—Doctoral Training Centre
16 Richmond Street, Glasgow G1 1XQ
Scotland, United Kingdom
44 (0)141 552 4400

The overall aim of the Research Centre is to meet the needs of the fast growing wind energy industry by providing high-caliber PhD graduates with the specialist and leadership skills necessary to lead future developments in wind energy systems. The objectives are to ensure that students from different disciplines gain competencies in core aspects of wind energy systems engineering.

University of Wisconsin

sel.me.wisc.edu
Solar Energy Program
1343 Engineering Research Bldg.
1500 Engineering Drive
Madison, WI 53706-1687
(608) 263-5626

For study at the Solar Energy Laboratory, which offers MS and PhD degrees, applicants must have a bachelor's degree in engineering or its physical science equivalent, have a strong undergraduate record and be admitted to either the mechanical or chemical engineering department. Three letters of recommendation are required. The GRE examination is recommended.

University of Wyoming

www.uwyo.edu
School of Environment and Natural Resources
P.O. Box 3963
Laramie, WY 82701-3963
(307) 766-5080

The University of Wyoming School of Environment and Natural Resources provides degree programs in research and policy aspects of environmental and natural resource decision-making. University of Wyoming focuses on six operational clusters: energy; ecosystems; allocation systems; environmental conditions and economic and social interactions; land and water resources; and atmospheric resources and systems. Additionally, the University's School of Energy Resources provides technical education in both renewable and nonrenewable energy regimes.

Vermilion Community College

www.vcc.edu
1900 East Camp Street
Ely, MN 55731
(800) 657-3608

VCC offers a two-year associate's degree in watershed science. Watershed science graduates work for either government agencies (like the USGS and local Soil and Water Districts) or for private companies that collect water information for profit. As hydrologic technicians, they measure stream flows, collect and analyze samples, monitor groundwater, and study the effects of human and natural activities.

Walla Walla Community College

www.wwcc.edu
Agriculture Center of Excellence
500 Tausick Way
Walla Walla, WA 99362
(509) 522-2500

The Bioenergy Department offers an associate's degree in bioenergy operations (see the text).

Washington State University

www.energy.wsu.edu
Energy Program at Olympia
905 Plum St. SE
Olympia, WA 98504-3165
(360) 956-2000

The Energy Program teaches energy and ventilation codes for construction staff and designers. The WSU web site offers extensive information on technical and trades-related issues.

Yavapai College

yc.edu
Construction Technology Department
2275 Old Home Manor Way
Chino Valley, AZ 86323
(928) 717-7720

Students learn through hands-on design and construction of a high-performance house using renewable energy. The college offers a one-year advanced certificate or a two-year AAS degree in residential building technology.

General Reference Web Sites

Additional details about these web sites can be found at the end of various chapters.

Association of Private Sector
Colleges and Universities
www.career.org

College Source Online
www.collegesource.org

Energy Management Institute
www.energyinstitution.org

Findsolar.com
www.findsolar.com

Florida Solar Energy Center
www.fsec.ucf.edu

National Energy Education
Development Project (NEED)
www.need.org

National Energy Foundation
www.nef1.org

National Fuel Cell Research Center
www.nfcrc.uci.edu

Natural Resources Defense Council
www.nrdc.org/energy

North American Board of Certified
Energy Practitioners
www.nabcep.org

Ocean Energy, California Energy
Commission
www.energy.ca.gov/oceanenergy/

Renewable Energy Access
www.renewableenergyaccess.com

Rocky Mountain Institute
www.rmi.org

SustainableBusiness.com
www.sustainablebusiness.com

Government Web Sites

Additional details about these web sites can be found at the end of various chapters.

Energy Information Administration
www.eia.doe.gov

Energy Star Program
www.energystar.gov

National Renewable Energy Laboratory
Advanced Vehicles & Fuels, Biomass, Buildings, Geothermal, Hydrogen & Fuel Cells, Solar, Wind
www.nrel.gov

National Renewable Energy Laboratory: Education
www.nrel.gov/education

Occupational Outlook Handbook
www.bls.gov/ooh
For an idea of current salaries in various fields connected to renewable energy, please consult this U.S. Bureau of Labor Statistics publication.

U.S. Department of Energy
Energy Efficiency and Renewable Energy
Biomass, Building Technologies, Federal Energy Management, Geothermal, Hydrogen & Fuel Cells, Solar Energy, Vehicle Technologies, Wind & Hydropower
eere.energy.gov

U.S. Environmental Protection Agency
www.epa.gov/opptintr/greenbuilding

Online Job Listings, & Hands-On Training

Additional details about these web sites can be found at the end of various chapters.

Dayaway Careers
www.dayawaycareers.com

Environmental Career Center
www.environmentalcareer.com

Environmental Career Opportunities
www.ecojobs.com

Environmental Careers Organization
www.eco.org

Green Energy Jobs
www.greenenergyjobs.com

Greenjobs
www.greenjobs.com

Midwest Renewable Energy Association
www.midwestrenew.org

Renewable Energy Access
www.renewableenergyaccess.com

Solar Energy International
www.solarenergy.org

Solar Living Institute
www.solarliving.org

SustainableBusiness.com
www.sustainablebusiness.com

U.S. Department of Agriculture; Careers with the Agricultural Research Service
www.ars.usda.gov/Careers/Careers.htm

Yestermorrow Design/Build School
www.yestermorrow.org

NOTE: Many professional societies and associations post jobs, as do major employers within each field. Check their web sites for details.

Professional Associations & Societies

Additional details about these web sites can be found at the end of various chapters.

American Institute of Architects
www.aiaonline.com

American Bioenergy Association
www.biomass.org

American Hydrogen Association
www.clean-air.org

American Society of Heating, Refrigeration and A/C Engineers
www.ashrae.org

American Solar Energy Society
www.ases.org

American Wind Energy Association
www.awea.org

Association of Energy Engineers
www.aeecenter.org

Electric Drive Transportation Association
www.electricdrive.org

Energy and Environmental Building Association
www.eeba.org

European Wind Energy Association
www.ewea.org

Geothermal Energy Association
www.geo-energy.org

Geothermal Resources Council
www.geothermal.org

International Association for Hydrogen Energy
www.iahe.org

International Solar Energy Society
www.ises.org

National Association of Home Builders
www.nahb.org

National Hydrogen Association
www.hydrogenassociation.org

National Hydropower Association
www.hydro.org

National Science Teachers Association
www.nsta.org

National Society of Professional Engineers
www.nspe.org

Solar Energy Industries Association
www.seia.org

Sustainable Buildings Industry Council
www.SBICouncil.org

U.S. Association for Energy Economics
www.usaee.org

U.S. Green Building Council
www.usgbc.org

State Energy Offices

Alabama Energy Division
Call: (334) 242-5290
www.adeca.alabama.gov/Energy

Alaska Energy Authority
Call: (907) 771-3000
Toll Free in AK: (888) 300-8534
www.akenergyauthority.org

Arizona Energy Office
Call: (602) 771-1137
www.azenergy.gov

Arkansas Energy Office
Call: (800) 558-2633
www.arkansasenergy.org

California Energy Commission Renewable Energy Programs
Call: (916) 654-4058
Toll Free in CA (800) 555-7794
www.energy.ca.gov/renewables

Colorado Energy Office Recharge Colorado
Call: (303) 866-2100
Toll Free: (800) 632-6662
www.rechargecolorado.com

Connecticut Energy Office
Call: (860) 418-6200
Toll Free: (800) 286-2214
www.ct.gov/opm

Delaware Division of Energy & Climate
Call: (302) 735-3480
www.dnrec.delaware.gov/energy

District of Columbia Energy Office
Call: (202) 673-6700
www.ddoe.dc.gov

Florida Office of Energy
Call: (850) 487-3800
www.freshfromflorida.com/Divisions-Offices/Energy/Office-of-Energy

Georgia Environmental Finance Authority – Energy Resources
Call: (404) 584-1000
gefa.georgia.gov/energy-resources

Hawaii State Energy Office
Call: (808) 587-3807
energy.hawaii.gov/

Idaho Office of Energy Resources
Call: (208) 332-1660
www.energy.idaho.gov

Illinois Energy Office
Call: (217) 785-3416
www.illinois.gov
(Search for 'renewable energy')

Indiana Office of Energy Development
Call: (317) 232-8939
www.in.gov/oed

Iowa Office of Energy Independence
Call: (515) 725-0431
http://iowaeconomicdevelopment.com/Programs/Energy

Kansas Energy Division
Call: (785) 271-3100
www.kcc.state.ks.us/energy

Kentucky Dept. for Energy Development and Independence
Call: (502) 564-7192
Toll-free in KY: (800) 282-0868
www.energy.ky.gov

Louisiana Energy Office
Call: (225) 342-1399
www.dnr.louisiana.gov
(Look under Energy)

Efficiency Maine
Call: (866) 376-2463
www.efficiencymaine.com

Maryland Energy Office
Call: (410) 260-7655
Toll Free: (800) 72-ENERGY
http://energy.maryland.gov/

Massachusetts Department of Energy Resources
Call: (617) 626-7300
www.magnet.state.ma.us/doer

Michigan Energy Office
Call: (517) 241-6228
www.michigan.gov/

Minnesota Office of Energy Security
Call: (651) 296-5175
Toll free in MN: (800) 657-3710
http://mn.gov/commerce/energy/

Mississippi Energy Division
Call: (601) 359-6600
Toll free: (800) 222-8311
www.mississippi.org/energy/

Missouri Division of Energy
Call: (573) 751-3443
Toll free: (800) 361-4827
http://ded.mo.gov/division-of-energy/renewables

Montana Energy Office Energize Montana
Call: (406) 841-5200
www.deq.mt.gov/energy

Nebraska Energy Office
Call: (402) 471-2867
Toll Free: (877) 337-3463
www.neo.ne.gov

Nevada State Office of Energy
Call: (775) 687-1850
www.energy.state.nv.us

New Hampshire Office of Energy and Planning
Call: (603) 271-2155
www.nh.gov/oep

New Jersey's Clean Energy Program
Call: (609) 777-3300
Toll Free: (866) 657-6278
www.njcleanenergy.com

New Mexico Energy Conservation and Management Division
Call: (505) 476-3310
www.emnrd.state.nm.us/ecmd

New York State Energy Research and Development Authority
Call: (518) 862-1090
Toll free: (866) NYSERDA
www.nyserda.org

North Carolina State Energy Office
Call: (919) 733-2230
Toll free in NC: (800) 662-7131
www.energync.net

North Dakota Office of Renewable Energy & Energy Efficiency
Call: (701) 328-5300
www.communityservices.nd.gov/energy

Ohio Advanced Energy and Efficiency Programs
Call: (614) 466-6797
http://development.ohio.gov/bs/bs_renewenergy.htm

Oklahoma State Energy Office
Call: (405) 815-6552
Toll free: (800) 879-6552
www.okcommerce.gov/State-Energy-Office/SEO-Overview

Oregon Department of Energy
Call: (503) 378-4040
Toll free: (800) 221-8035
www.oregon.gov/energy

Pennsylvania Office of Energy & Technology Deployment
Call: (717) 783-0540
www.depweb.state.pa.us/energy

Rhode Island Office of Energy Resources
Call: (401) 574-9100
www.energy.ri.gov

South Carolina Energy Office
Call: (803) 737-8030
Toll free: (800) 851-8899
www.energy.sc.gov

South Dakota Energy Management Office
Call: (605) 773-3899
www.state.sd.us/boa/ose/OSE_ Statewide_Energy.htm

Tennessee Office of Energy Programs
Call: (615) 741-2994
tn.gov/environment/energy.shtml

Texas State Energy Conservation Office
Call: (512) 463-1931
www.seco.cpa.state.tx.us

Utah State Office of Energy Development
Call: (801) 537-3300
http://energy.utah.gov/renewable-energy/

Vermont State Energy Office
Call: (802) 828-2811
Toll free: (800) 622-4496
www.publicservice.vermont.gov
(Look under Renewables Energy, and Energy Efficiency)

Virginia Division of Energy
Call: (804) 692-3200
www.dmme.virginia.gov/DE/ DeLandingPage.shtml

Washington State Energy Office
Call: (360) 725-3118
http://www.commerce.wa.gov/ Programs/Energy/Office/

West Virginia Division of Energy
Call: (304) 558-2234
Toll free: (800) 982-3386
http://www.wvcommerce.org/info/ aboutcommerce/energy/

Wisconsin Office of Energy Independence
Call: (608) 261-6609
www.energyindependence.wi.gov

Wyoming State Energy Office
Call: (307) 777-2800
Toll free: (800) 262-3425
www.wyomingbusiness.org/energy

Glossary

ABET. Accreditation Board for Engineering and Technology (ABET), the recognized accreditor for college and university programs in applied science, computing, engineering, and technology, made up of 28 professional and technical societies representing these fields.

AC. Alternating current; a form of electricity that cycles from minimum to maximum and then decreases to minimum again. The cycle is then repeated, now in the opposite direction. In the United States, alternating current changes direction sixty times per second.

alternative energy. Energy derived from nontraditional sources such as compressed natural gas, solar panels, hydroelectric turbines, wind turbines, and the like; such energy is alternative relative to fossil fuels, but not all forms of alternative energy (such as nuclear power) are renewable.

apprenticeship. The chief means of crafts instruction before the rise of technical and trade schools, apprenticeship is a system by which a novice serves as an assistant to a skilled worker, learning techniques and practices. Today, apprenticeship typically combines on-the-job training with supplemental reading or coursework.

architect. In the United States and Canada, a trained, accredited professional who designs buildings and often oversees their construction.

automotive designer. A specialist in the design of automobiles and other vehicles.

biobutanol. A noncorrosive combustible alcohol derived from biomass and used, among other purposes, as a fuel.

biochemist. A scientist who studies the chemical processes of living things, such as the production of sugars in plants.

biodiesel. Ester obtained by reacting methanol with vegetable oil, used in pure form or blended with diesel. Rapeseed, sunflower, palm, soybean, and peanut oils are most commonly used in biodiesel preparations.

bioenergy. Fuel or electricity derived from biomass. See also ethanol.

bioethanol. Ethyl alcohol derived from the fermentation of sugar from wheat, sugar beets, corn, and other plant sources.

biofuel. See *bioenergy*.

biogas. A liquid or gaseous fuel made from biomass. Some biogas is made by recovering methane from landfills, stockyards, and similar venues.

biologist. A scientist who studies living organisms and their relationship to the environment.

biomass. Organic matter available on a permanent, renewable, and sustainable basis, such as trees, agricultural crops, and waste (such as straw, wheat stalks, and pulp). The complex carbohydrates in such matter is converted to biogas and other forms of bioenergy.

blue-collar. A sociological term used to designate skilled and unskilled workers in industrial or manual jobs, so called because of the coarse blue shirts such workers wore, in contrast to administrative and managerial (white-collar) jobs.

Btu. British thermal unit; a unit of thermal energy used in reference to heating or cooling. The amount of energy required to raise the temperature of a pound of water by one degree Fahrenheit.

carbon-neutral. Carbon neutrality means balancing the amount of carbon dioxide released into the atmosphere by sequestering an equal amount. In principle, carbon neutrality is difficult for an individual to achieve; reducing CO_2 emissions through conservation has proven the more efficient tactic.

cellulosic biomass. The fibrous, woody, usually inedible portions of plants, comprising about 75 percent of all plant material and used to make biofuel.

chemical engineer. An engineer who specializes in using chemical processes in industrial uses, as in the production of biofuels from cellulosic materials.

CHP. Combined heat and power; the generation of electrical power and usable heat from a combustion process.

civil / structural engineer. An engineer who specializes in the construction of built structures such as dams, bridges, roads, canals, and buildings, in the latter often working closely with an architect.

DC. Direct current, a form of electricity in which the current moves in only one direction, unlike alternating current.

electric car. Automobile powered by electricity, rather than by petroleum or other liquid fuel.

EMCS. Energy Management Control Systems; tools that promotes energy efficiency by accurately monitoring energy consumption and building operations.

electrical engineer. An engineer who specializes in the design, construction, and maintenance of electrical systems.

energy analyst. A trained specialist charged with studying the uses and costs of energy in buildings, manufacturing processes, and the like, usually with an eye to improving efficiencies.

energy audit. An inspection process that determines how much energy is used, usually accompanied by suggestions for conserving energy.

energy economics. A broad subject area which includes a variety of topics related to the supply and use of energy in our societies. Economists study the ways in which we allocate our limited energy resources to meet our unlimited needs / wants.

energy efficiency. An engineering concept whereby machines, structures, manufacturing processes, and the like are designed or revised to decrease the amount of energy used, the governing idea being to improve efficiency rather than have to generate more power.

energy manager. Within industry, an engineer or trade worker whose principal task is to conduct an energy audit, identify ways to improve energy efficiency, and the like.

environmental engineer. An engineer who specializes in applying scientific and engineering principles to improving the state of the environment, often by remediating polluted sites, and who helps identify sites for the construction of utilities, roads, and other such structures.

ethanol. Biofuel derived from grain, corn, sugarcane, or other starchy plant material and used alone or as an additive to gasoline.

fenestrations. Glazed aperatures in buildings, including windows, glass doors, clerestories, and skylights.

fossil fuels. Carbon- and hydrogen-laden fuels formed underground from the remains of long-dead plants and animals, such as crude oil, natural gas and coal.

FSEC. Florida Solar Energy Center; a research institute of the University of Central Florida.

fuel cell. An electrochemical device that converts chemical energy to electrical energy without combustion.

fuel engineer. An engineer who specializes in the production of fuels of various kinds.

geochemist. A scientist who specializes in the study of the chemical composition of the earth and other planets.

geoexchange system. An electrically powered heating and cooling system for interior spaces, using the earth as a heat source and heat sink.

geologist. A scientist who specializes in the study of the origins, composition, and physical processes of the earth and other planets.

geophysicist. A scientist who specializes in the study of the physics of the earth, particularly its electrical, gravitational, and magnetic fields.

geothermal energy. Heat generated by natural processes within the earth. The chief energy resources are hot dry rock, magma (molten rock), hydrothermal (water/steam from geysers and fissures), and geopressure (water saturated with methane under tremendous pressure at great depths). Such heat is typically brought up through pipes with the use of a geoexchange system.

GHP. Geothermal heat pump; a heat pump that uses the earth as a heat source and heat sink.

gigawatt (gW). Unit of energy equivalent to one billion watts or one million kilowatts.

green building. A building or construction method that makes use of sustainable, renewable materials and principles and strives for energy efficiency.

green transportation. Vehicles such as bicycles or fuel cell–powered cars that do not require the burning of fossil fuels for their power.

green-collar. A play on words of the old designations blue-collar and white-collar. Refers to jobs in the "green" field of renewable energy.

HVAC. Heating, ventilation, and air conditioning, an acronym used to describe such systems and the engineers who work with them.

hybrid car. A vehicle that uses both an on-board rechargeable battery system and a fuel-based power source for propulsion. Most hybrid vehicles recharge their batteries by capturing kinetic energy from braking, with fuel used either to generate more electricity through a generator or to provide extra power when it is needed.

hydraulic engineer. A civil engineer who specializes in the study of moving fluids, particularly in how to control that flow and convey it to places where it is wanted.

hydroelectric power. Electrical power generated by the kinetic energy of moving water, often at a natural or artificial waterfall or in a swiftly moving river, but increasingly in oceanic environments through the harnessing of wave energy.

hydrogen energy. Energy derived from hydrogen, a renewable and emission-free fuel, now largely applied to automobiles.

hydrologist. A scientist who studies the distribution of water in the environment, often helping determine where to place dams, wells, and other structures.

hydropower. See *hydroelectric power* and *wave energy*.

IBEW. International Brotherhood of Electrical Workers

information technology (IT). Deals with the use of electronic computers and computer software to convert, store, protect, process, transmit and retrieve information, securely.

kilowatt (kW). Unit of electrical power equal to one thousand watts.

LEED. Leadership in Energy and Environmental Design; a nonprofit organization that has established the Green Building Rating System and other guidelines in sustainable building. See *www.usgbc.org*

life-cycle engineering. An emerging research field closely related to sustainable development. LCE comprises assessment and costing models that aim at cost control, efficiency, and optimization. LCE defines the life cycle of a product from raw material through its use, disposal, and, ideally, reuse.

marine energy. See *wave energy.*

material scientist. A scientist or engineer who specializes in the study of matter and its properties and applications in industrial uses.

mechanical engineer. An engineer who specializes in the design, construction, and maintenance of mechanical systems, such as a power plant.

megawatt (MW). One million watts, or a thousand kilowatts.

meteorologist. A scientist who specializes in the study of climate and weather systems, often with an interest in forecasting near-term conditions.

methane recovery. See biogas.

microbiologist. A biologist specializing in the study of microorganisms, such as protozoa, fungi, bacteria, and viruses.

micro-hydro. Small-scale hydropower, producing a hundred kilowatts or less.

NABCEP. North American Board of Certified Energy Practitioners, an association that offers training and certification to renewable-energy professionals.

NREL. National Renewable Energy Laboratory; a research facility run by the U.S. Department of Energy in Golden, Colorado, devoted to all forms of renewable energy.

oceanographer. A scientist whose area of study is the ocean and its environments and inhabitants.

PE. Professional Engineer; in the United States and Canada, a registered or licensed engineer authorized to offer professional services directly to the public. A PE must be a graduate of an engineering program accredited by ABET and have passed several rigorous examinations, among other requirements.

photovoltaic (PV). The technology of converting sunlight directly into electricity through the use of photovoltaic (solar) cells. Photovoltaic cells are made primarily of silicon, which responds electrically to the presence of sunlight, a phenomenon known as the photovoltaic effect.

process engineer. An engineer. who develops industrial processes that are in turn used to make other things. Most process engineers hold degrees in chemical or mechanical engineering and have extensive knowledge in how a chemical or manufacturing plant works.

project manager. In many trades such as construction, the person in charge of establishing and maintaining schedules, budgets and payroll, staffing, and the like.

quadrillion. 1 quadrillion Btu equals 1,000,000,000,000,000 Btu (British Thermal Unit).

RA. Registered Architect, certification earned through the National Council of Architectural Registration Boards following graduation from an accredited architecture school by fulfilling a number of educational and work requirements.

renewable energy. Any form of energy that derives from sustainable resources that can be naturally renewed—wind, for instance, or energy from the sun. A general way to think of renewable energy is any energy whose production does not require burning anything, which distinguishes it from alternative energy.

soil scientist. A scientist who studies the chemistry and organic composition of the soil, principally in order to determine the suitability of a given plot of land to agricultural production, but also to determine siting for energy facilities.

solar collectors. Devices for capturing the sun's energy; flat-plate collectors and evacuated solar tubes are two types.

solar electricity. Electricity generated directly by sunlight via photovoltaic cells.

solar energy. See *solar electricity* and *solar thermal.*

solar engineer. A trained person who designs solar energy systems (solar electric systems as well as solar thermal systems).

solar installer. A trained craftsperson who specializes in installing and maintaining photovoltaic assemblies and other solar energy–related equipment.

solar thermal. Heat, rather than electricity, that is generated by the sun and put to a useful purpose, such as for home heating and water heating.

solar water heating. See *solar thermal.*

Stirling engine. A closed-cycle external combustion engine in which heat is applied through the wall of a chamber within which a gas is heated and cooled, thus expanding and contracting to power a piston. Stirling engines are quiet and highly efficient, but also expensive.

sustainable building. Construction that uses renewable materials and is energy-efficient, sometimes to the point of being carbon-neutral.

tidal energy. See *wave energy.*

turbine. A machine with propeller-like blades, or rotors, that are moved by flowing water, air, or gas to power a generator, thereby producing electricity.

utility grid. The system of electrical utilities, tied in North America into a grid such that electricity generated in California can, in theory, be transmitted to Alaska or Florida.

watt (W). Unit of electrical power used to indicate the rate of energy produced or consumed by an electrical device.

wave energy. Technically, a form of solar energy, whereby ocean waves are generated by solar-driven winds. Systems are in development to make use of this clean and endlessly renewable source of energy.

white-collar job. In the U.S., a white-collar worker performs professional, managerial, or administrative work. Other types of work are those of a blue-collar worker, whose job requires manual labor. See also *blue-collar.*

wind energy. Electrical power derived from moving air via a turbine that drives a generator.

Index

Numbers shown in italics represent photos /illustrations

5/23/17
20,00

The Author

G regory McNamee is a writer, editor, photographer, and consultant in publishing, film, and other media. He is the author or title-page editor of thirty-six books, including several on ecology and natural history, and of more than four thousand periodical pieces. He writes on science, culture, the environment, and other topics for many publications in the United States and abroad, and he is a contributing editor to the Encyclopaedia Britannica. McNamee divides his time between his home in Tucson, Arizona, and his migratory habitats in Virginia, New York, Italy, and other places.

Careersinrenewableenergy.com

Check out McNamee's blog:
careersinrenewableenergy.wordpress.com

100% solar & wind powered since 1999

PIXYJACK PRESS INC

Wholesale Orders Welcome
PO Box 149, Masonville, CO 80541 USA
PixyJackPress.com *info@pixyjackpress.com*